A SURVIVAL GUIDE TO THE MISINFORMATION AGE

DAVID J. HELFAND

A SURVIVAL GUIDE TO THE MISINFORMATION AGE

scientific habits of mind

Columbia University Press   New York

Columbia University Press
Publishers Since 1893
New York   Chichester, West Sussex
cup.columbia.edu
Copyright © 2016 David J. Helfand

Library of Congress Cataloging-in-Publication Data
Helfand, D. J. (David J.), 1950–
    A survival guide to the misinformation age : scientific habits of mind /
David J. Helfand.
        pages cm
    Includes bibliographical references and index.
    ISBN 978-0-231-16872-4 (cloth : alk. paper) — ISBN 978-0-231-54102-2 (ebook)
1. Errors, Scientific. 2. Science—Methodology. 3. Statistics—Methodology.
4. Missing observations (Statistics) I. Title.

Q172.5.E77H45 2016
500—dc23

2015034152

Jacket design by Alex Camlin

CONTENTS

CONTENTS

FOREWORD

For nearly a century, the undergraduate college of Columbia University has required all first- and second-year students to engage in discussion and contemplation of some of the great ideas that Western Civilization has produced. For most of that period, the seminal works for these courses have been drawn exclusively from the humanistic tradition. Since 1937, Columbia's Core Curriculum has consisted of seven courses that cover the intellectual and cultural history of the West through the study of literature, political philosophy, music, and art. Although the Core was described as the "intellectual coats of arms" of the university, science and mathematics were absent. In 1982, I chaired a committee that recommended this lacuna be rectified by adding to the Core a course in science. Twenty-two years later, in the university's 250th year, Frontiers of Science was launched as a class required of all first-year Columbia College students. In 2013, the idea that science should become a permanent component of the Core Curriculum was formally adopted.

Unlike most general education and introductory-level university science courses, Frontiers of Science was not designed to impart to our students large quantities of information about a particular scientific field—the products of scientific inquiry. Rather, the point was to use examples drawn from a number of disciplines to illustrate what science is and how it produces our understanding of the material universe. I developed a brief, online text called "Scientific Habits of Mind" in conjunction with Frontiers of Science as an introduction to the distinct modes of thought that scientists use in producing their unique and powerful models of the world. Habits form the core of this book.

Over the thirty-eight years I have taught at Columbia, the College has vastly expanded its applicant pool and has significantly improved its selectivity. Yet I have found that among the students now arriving on campus, preparation in basic quantitative reasoning skills has declined at an alarming rate. In the 1970s, many of our students had not taken calculus in high school; now the vast majority of those we admit have had at least one calculus course. But their ability to use numbers, read graphs, understand basic probability, and distinguish sense from nonsense has declined. And in the larger population—among politicians, journalists, doctors, bureaucrats, and voters—the ability to reason quantitatively has largely vanished. This is scary.

We live in a world dominated by science and its product: technology. This world faces daunting challenges—from energy supplies to food supplies, from biodiversity collapse to the freshwater crisis, and, at the root of many of these issues, global climate change. Yet we shrink from confronting these challenges because we don't like numbers and are more comfortable with beliefs than with rational thought.

Faith will not quell the increasing demands humans place on the Earth's resources. New Age thinking will not produce "sustainable development." It is not clear that any philosophy will allow this planet to sustain a population of ten billion people in the manner to which contemporary Western societies have become accustomed. But it is clear that, to assess the limits the Earth imposes, to contemplate rationally what routes we might take, and to plan a future free from war and want, quantitative reasoning must be employed. Blind faith—in God during the Middle Ages, in creativity during the Renaissance, in reason during the Enlightenment, and in technology today—is a shibboleth. It is a fantasy ungrounded from a rational assessment, in quantitative terms, of what is possible and what is not.

This book seeks to provide a set of tools to be used in promoting a rational and attainable future for humankind. It offers no formula for financial success or any promise of a sleeker physique. Rather, its goals are to cultivate rational habits of mind and to provide warnings against those who would pervert this uniquely human capacity so that our species might accommodate itself to the web of life that has been evolving here for 3.8 billion years—to allow that life, and the fragile intelligence it has produced, to remain a lasting feature of our planet.

ACKNOWLEDGMENTS

I arrived at Amherst College in the autumn of 1968 fully intending to major in theater studies. My wife, an accomplished actress on Broadway and television, assures me now that my continued pursuit of that path would have led to a lifetime of waiting tables. Thus, my first acknowledgment goes to the late Professor Waltraut Seitter for redirecting me toward astrophysics. I wasn't averse to science in school, but I didn't see how anyone could love it. Professor Seitter showed me the way. The first woman to obtain a chair in astronomy in Germany, Professor Seitter was teaching at Smith College in 1969 when I accidentally encountered her in a class I never planned on taking. Such is the role of serendipity in life, and I am grateful for it.

My physics and astronomy professors at Amherst (Joel Gordon, Bob Romer, George Greenstein, and Skip Dempsey—whom I was delighted to encounter recently at one of my public lectures) built my foundation in science. My many graduate mentors at the University of Massachusetts ingrained in me the habits of mind that have served me well and form the basis of this book; most noteworthy are my undergraduate thesis advisor Dick Manchester and PhD supervisor Joe Taylor.

ACKNOWLEDGMENTS

Colleagues at Columbia and astronomical collaborators from around the world too numerous to mention here have, over the past four decades, also helped to shape my views of science. Those who participated in my quixotic quest to add science to Columbia's Core Curriculum, the ultimate stimulus for this book, however, deserve special thanks. Dean Bob Pollack was steadfast in championing this idea through more than one failed attempt, and Provost Jonathan Cole provided key support fifteen years ago when the latest push finally came to fruition. That would not have happened without my compatriots Jacqueline van Gorkom and Darcy Kelley, who were the earliest participants, or without the many faculty from diverse departments who agreed to join the enterprise: Don Hood, Nick Christie-Blick, Don Melnick, Wally Broecker, and Horst Stormer deserve special mention. Elina Yuffa has been holding the course together since its inception, and the first short, online version of this text would never have been completed without the tireless work of Ryan Kelsey at the Columbia Center for New Media Teaching and Learning and the creative assistance of Eve Armstrong, some of whose words and images remain in the boxes and endnotes of this volume. Special thanks—and absolution from any errors that may have crept into chapters 7 and 8—is owed to Columbia statistician Dan Rabinowitz, who greatly clarified some of my amateur statistical reasoning. Quest students Morgaine Trine, Annie Borch, and Camilo Romero were invaluable in the final push to complete the manuscript.

My editor Patrick Fitzgerald at Columbia University Press was persistent to a fault in convincing me to finally sit down and get this done; he was overwhelmingly generous with advice, deadlines, and the myriad other details with which first-time authors are rarely familiar. Kate Schell ably assisted.

It is, appropriately, Valentine's Day as I write this, since my final acknowledgment goes to my actress spouse who has, retroactively, applauded so convincingly my switch from theatre to science. Whether it has been to amuse, to bemuse, or to be a muse, you are always there, Jada, and for that I am unstintingly grateful.

A SURVIVAL GUIDE TO THE MISINFORMATION AGE

INTRODUCTION

Information, Misinformation, and Our Planet's Future

Throughout human history, information has been limited, difficult to access, and expensive. Consider the decades-long apprenticeship required to become the shaman in a hunter-gatherer tribe. Or picture a medieval monk in his cell, copying by hand a manuscript from Aristotle for the monastery's library—a place where only the monks could browse. By 1973, when I published my first scientific paper, little had changed. *The Astrophysical Journal,* the leading scholarly publication in my field, included a total of 10,700 pages that year (a lot to thumb through looking for a specific piece of information) and cost $651 for an annual subscription (in inflation-corrected dollars)—difficult to access, expensive, and largely limited to American astronomers.

In the last decade, this situation has been radically reversed: information is now virtually unlimited, ubiquitous, and free. Google the "complete works of Aristotle," for example; within one second you can find http://classics.mit.edu/Browse/browse-Aristotle.html and download—for free—every surviving word that Aristotle wrote. And in my field? Virtually every scientific paper on astronomy published in the world since the nineteenth century is instantly available at http://adsabs.harvard.edu/abstract_service.html, where you can search for any word and for every mention of every celestial object ever named; you can search by author, journal, title, which paper another paper was cited in, etc. And you can do this anywhere on your smart phone—again, at no charge whatsoever. Unlimited, ubiquitous, and free.

This epochal transformation, however, is not an unbridled blessing. According to a study published a couple of years ago by IBM,[1] with the proliferation of electronic communication devices and services— from tablets to smart phones and from social media sites to e-mail— we are now creating 2.5 quintillion bytes of new data every day (that's 2,500,000,000,000,000,000 bytes or the equivalent of five *trillion* books the length of this one, enough to fill bookshelves half a kilometer high stretching around the Earth at the equator). The IBM report goes on to note that roughly 90 percent of all the information that exists in the world today was created in the past two years. Not much time for editing, testing, and serious reflection.

As a result, accompanying Aristotle and the *Astrophysical Journal* on the Web is a tsunami of misinformation that threatens to overwhelm rational discourse. Much has been written about the democratizing power of the Web; less has been said about its virtually unlimited capacity for spreading misinformation—and disinformation—in the service of ends that are distinctly undemocratic and unhealthy both for individuals and for our planet.

Why is misinformation so prevalent?

Information is transferred when the thought of a single mind is shared with another. In hunter-gather societies, information may have been very limited, but it was highly accurate and self-correcting—misinformation was minimized. A hunter who routinely led his kin away from the best hunting grounds was soon either ostracized or simply ignored. Anyone who gathered poisonous berries instead of nutritious ones was quickly eliminated from the gene pool.

In his recent book *The Social Conquest of Earth*,[2] E. O. Wilson describes in evolutionary terms how this sharing of valid information was rewarded. My favorite theorem from this book states that whereas a selfish individual will always outcompete an altruistic individual, a group of altruists will always outcompete a group of selfish individuals. Wilson's view is that collaboration among humans has become an essential part of our wiring through natural selection. Throughout most of human prehistory, then, validating information and sharing it was both valued and rewarded. Information may have been quite limited, but its quality was high.

We now occupy the opposite state: information is virtually unlimited but often of very low reliability. How did this happen?

When one doesn't know, has never met, and, furthermore, shares no common interest with the individual with whom information is being shared, there is little incentive to assure the reliability of that information. Indeed, if there is personal gain to be had by providing misinformation to one's unknown targets, there is, in fact, an incentive to misinform. Wilson's altruism is claimed to have evolved in groups of approximately thirty. (Although no evidence is adduced for this number, it is clearly not wrong by a factor of ten in either direction.) However, in a target market of seventy million consumers—or seven billion—there is no incentive for altruism. If more profit—in dollars, in power, in fame—is to be gained from misinformation, why not go for it? And when dollars, power, and fame are combined (as they often are in politics), the incentive is all the stronger. Misinformation predominates.

This book is designed as an antidote to the misinformation glut.

You are living with a brain that evolved over several million years to make simple decisions on the plains of the Serengeti: find food, avoid becoming food, and reproduce. You are living in a fast-paced, rapidly changing world dominated by modern technology. This is a challenging mismatch.

The primary drives are still there, of course, but your means of fulfilling them have altered radically. Find food: a drive-through McDonalds or a stroll through Whole Foods—no spears or snares required. Avoid becoming food: abjure climbing the fences of the zoo's lion enclosure. Reproduce: Match.com or the in vitro fertilization clinic.

The number and range of decisions you must make to satisfy these simple drives have expanded greatly. And then there are the decisions our ancestors hadn't imagined: equity mutual funds or corporate bonds for the retirement account? Radiation or chemotherapy? Waldorf or International Baccalaureate? Tea Party or Democrat? The Serengeti operating system just isn't much help. You need to invoke the new processor (the prefrontal cortex) that works more slowly but can handle more complex situations—if it has all the right apps installed. I collect those apps here: I call them scientific habits of mind.

To make wise decisions concerning finances, health, education, and politics—to even formulate well-informed opinions about these and other issues—you must own the tools required to evaluate, and validate, information for yourself. No search engine can do this for you, and relying on appeals to authority can get you into a lot of trouble. Relying on personal anecdotes (your "experiences" or those of others) is even worse. The most effective approach to all problems requiring rational analysis is to cultivate scientific habits of mind.

Although curiosity—the impetus for science—has been a feature of humankind for at least several hundred thousand years, science is, evolutionarily speaking, an extremely recent invention. It is not, therefore instinctual, not intuitive. But it provides a very powerful model for the physical world. Science has also developed a set of tools, often, but not always, couched in mathematical terms, that allow us to evaluate information, place it in context, and deduce consequences—in short, to provide a rational basis for decisions that are in touch with the physical reality of the universe.

My goals here, then, are twofold: first, to offer a glimpse of the enriching experience science affords to those who grasp its workings, and second, to offer a guide for cultivating the habits of mind necessary for successfully navigating the modern world. Ironically, the technology spawned by science has unleashed upon the world a deluge of misinformation that has, in a decade, gone from being merely annoying to outright dangerous.

Some of this misinformation masquerades as science, adopting technical jargon, holding conferences, and publishing journals. Although it is politically incorrect to do so in some circles, I call this what it is: pseudoscience, which I define as a body of inquiry that superficially adopts the language and trappings of science but operates in a parallel universe, rejecting all findings that do not support its basic tenets, is immune to falsifiability, and has no impact on the ideas and work of those pursuing genuine scientific inquiry. Some sociologists of science will, no doubt, be inflamed by that last statement, citing it as prima facie evidence of manifest white, male, hegemonic arrogance on my part. So be it—they can write their own books.

We find ourselves at the start of the third millennium by our current method of calendrical reckoning, the 4,568,000th millennium calculated in geologic time. And this millennium is unique. Never before in Earth's history have there been seven billion large mammals occupying almost every ecological niche on the planet. More importantly, never before has a species possessed the ability to contemplate its future. Unfortunately, far too few of these seven billion mammals have the leisure time, the resources, or the inclination to think much about the future. The fate of our species is in the hands of those who do have this luxury. My goal is to equip you for this task by cultivating those habits of mind I believe are essential if our species is to thrive at the dawn of the fourth millennium.

1
A WALK IN THE PARK

Would you rather learn about stellar nucleosynthesis
or go for a walk in the park?

f I were to rank all of the questions I have asked in my thirty-eight years of teaching, from easiest to hardest, this would probably take the top spot.

It was the first perfect spring day in New York—68 degrees under a cloudless cerulean sky with a barely perceptible breeze flitting in off the Hudson River. My class of seventy Columbia College students had torn themselves away from an early afternoon idyll on the lawn and reluctantly slid into the hard wooden seats of my air-conditioned, artificially lit lecture hall at the appointed time of 2:40 P.M.

My planned lecture on how the stars cook up the atoms that compose our bones and flesh could not compete with my tempting—and most unexpected—offer. But there was a catch I told them. If we went for a walk in the park, they would have to hear how *I* experienced the walk—as a scientist. They were undeterred; the vote for walking was unanimous.

I led them down a flight of stairs and out through the loading dock onto 120th Street into the shadow of Pupin Labs. Why into the shadow? Because we were at 40.7 degrees north latitude and the Manhattan street grid runs[1] (not quite) east–west. The angle of the Sun above the horizon at

2:40 P.M. on a spring day is insufficient to clear the thirteen-story[2] building we had just exited.

"What do you feel on your face?" I asked. After a confused silence, one student said, "It feels nice."

"The reason it's nice," I replied, "is because the air molecules—nitrogen, oxygen, and other trace constituents—are moving at about 450 meters per second today and are colliding with your skin molecules billions of times per second. These collisions trigger protein interactions in the temperature-sensing nerve cells of your skin, which in turn send an electrical signal to your thalamus, informing your prefrontal cortex that this is a copacetic environment to hang out in—that is, it's pleasantly warm out here."

If some had thoughts of returning to stellar nucleosynthesis, they didn't let on, so we proceeded west on 120th and into Riverside Park.

Along the way, I solved the mystery of why the sky is blue (no, it is not a reflection of the color of the ocean—have you ever put seawater in a glass jar and discovered it has an azure hue?), why the pavement felt warmer than the air, why the Sun's disk could be covered by a fingernail held at arm's length, and why the paving stones of the park's sidewalk are hexagonal in shape. As we entered a wooded area of the park, I made an offhand comment that all the daffodils emerging from the leaf litter were a gift from the Dutch government, an offering of solace for a wounded city after the 9/11 attack.

"What daffodils?" "Where?"

I will admit the bulbs had yet to flower, but the unmistakable, straight green shafts were all around us. Sadly, few, if any, of the students recognized that a glorious carpet of yellow would grace this glen in a few days. They don't need a physicist to explain the subtleties of the unseen world around them, I thought; they need to start by regressing fifteen years or so, back to when they were curious about every aspect of the world and surely would have asked, "What are all those big fat blades of grass, Daddy?"

What are those big blades? *Why* are they so fat? *Where* do they come from? All questions one would expect from a five-year-old on a walk through the park. My etymological dictionary says these words—what, why, and where—come from Old English, Saxon, Proto-German, and Norse. I wonder, however, about their earlier origins and their relationship to the *whaaa* sounds that children make long before they can speak. It is an easy, natural sound—little more than an exhalation of breath, and much more natural than the voiced alveolar lateral continuant "l" followed by the long vowel phoneme "i" and the aspirated "k" that form "like," an utterance many eighteen-year-olds seem to feel is required between every fourth word in a sentence. I do often wish they'd go back to "why."

It is not the word choices that concerned me, however, as much as the disconnection from the natural world. I am not a fan of the proliferating number of "disorders" that we are told now afflict us, but the first two-thirds of the term *nature-deficit disorder*[3] seemed appropriate. How could they have lived eighteen years and never noticed those straight, thick blades as the harbingers of spring? But more importantly, where had their curiosity gone? Why *is* the sky blue, and why is it bright all over when the Sun just shines from one direction? Why does one's cheek feel warmer in the Sun than in the shade? Why are we enjoying this walk in the park more than the lecture on nucleosynthesis?

Why. It's such a satisfying word to utter and such a visceral pleasure when understanding emerges with the answer. The light from the sky opposite the Sun is still sunlight; it has just scattered off the molecules of air in such a way that these rays precisely target your pupil and no one else's. The wavelength of blue light is optimally tuned for this scattering, whereas the red, yellow, and green light passes directly through the atmosphere and thus appears to come only from the direction of the Sun—except when the Sun is near the horizon and the path the light must travel through the atmosphere is much longer, so that even some of the red and yellow light is scattered sideways, creating a memorable sunset (we'll discuss sunsets further in chapter 5).

The sunlight warms your cheek because, having been created through the fusion of protons in the Sun's core and having struggled 100,000 years to reach its surface, this light finally broke free and, after traveling in a perfectly straight line for eight minutes and nineteen seconds,[4] was absorbed by the molecules of your skin, which upped the tempo of their jiggling, a jiggling that was recorded in your brain as warmth. From protons fusing before civilization emerged to your glowing cheek today—it's a beautiful and compelling story.

I know—not everyone sees it this way. Consider, for example, these famous lines from British Romantic poet John Keats's 1820 narrative poem "Lamia":

> Do not all charms fly
> At the mere touch of cold philosophy?
> There was an awe-full rainbow once in heaven:
> We know her woof, her texture; she is given
> In the dull catalogue of common things.
> Philosophy will clip an Angel's wings,
> Conquer all mysteries by rule and line,
> Empty the haunted air, and gnomèd mine—
> Unweave a rainbow, as it erewhile made
> The tender-person'd Lamia melt into a shade.[5]

We are told by Margaret Robertson, a Keats scholar, that the damnation of this "cold philosophy"—such as "unweav[ing] the rainbow" by explaining it as light refracted in water droplets—is not to be taken as Keats's own view. Rather, it is the message of his allegorical poem that "it is fatal to attempt to separate the sensuous and emotional life from the life of reason."[6] Perhaps, but I am unconvinced.

That "all charms fly" under the onslaught of scientific reasoning was a common theme of the Romantic period. William Blake's famous monotype of Isaac Newton exudes a horrified rejection of the scientific mind that is expressed directly in Blake's aphorism: "Art is the tree of life. Science is the tree of death." The following poem by Edgar Allan Poe, "Sonnet—To Science,"[7] written in 1829 when Poe was twenty years old, is said to have been inspired by Keats's lines:

> Science! true daughter of Old Time thou art!
> Who alterest all things with thy peering eyes.
> Why preyest thou thus upon the poet's heart,
> Vulture, whose wings are dull realities?
> How should he love thee? or how deem thee wise,
> Who wouldst not leave him in his wandering
> To seek for treasure in the jewelled skies,
> Albeit he soared with an undaunted wing?
> Hast thou not dragged Diana from her car?
> And driven the Hamadryad from the wood
> To seek a shelter in some happier star?
> Hast thou not torn the Naiad from her flood,
> The Elfin from the green grass, and from me
> The summer dream beneath the tamarind tree?

I respectfully disagree with this Romantic view, so much in the ascendancy today. The wood and water nymphs Poe mourns (the Hamadryad and Naiad, respectively) may indeed be gone. Diana (the Moon), we now know, is not a radiant goddess but a globe of rock that was sheared off from the surface of the newly forming Earth 4.5 billion years ago and that keeps a regular appointment in the sky each night because it is constrained to roll around inside an imaginary bowl created by a subtle warping of spacetime induced by the presence of the nearby Earth. I don't find this a "dull reality" but a magnificent triumph of human curiosity, imagination, and analytical power.

This debate is not just a matter of differing tastes, however; it is now an existential matter for our species.

At the time Keats's "Lamia" was published, the world population had just exceeded one billion. Life expectancy in Europe and the United States was approximately thirty-eight years. It has doubled since, and the global population has increased more than sevenfold. It took roughly 150,000 years for *Homo sapiens* to generate a worldwide population of one billion; the last billion were added in just twelve years.[8] We have insinuated ourselves into almost every ecological niche available on the planet—regardless of whether we complement the diversity of existing life in each niche or not. We are using the demonstrably finite resources of the Earth at a manifestly unsustainable rate and are fundamentally altering the chemistry of the atmosphere and the oceans in a manner that is changing the energy balance of the planet.

A simple, quantitative estimate of our impact is instructive. An animal's basal metabolic rate is the sum of the energy consumed by the organism while it is at rest—the total energy required to maintain its body temperature and keep all its vital systems humming along. For an adult human this is roughly 100 watts. A watt is a measure of power—the rate at which energy is used. To stay alive, a person burns energy at the same rate as a 100-watt light bulb. All other animals, from cockroaches to elephants, consume energy and produce waste equivalent to their individual basal metabolic rates.

But humans are uniquely clever. We also use energy to ease our travel, communicate across vast distances, alter the temperature of our environments, and manufacture toys to entertain ourselves. In fact, on average, North Americans use energy not at the 100-watt rate of their hunter–gatherer ancestors but at 100 times this rate: 10,000 watts. As the climatologist Richard Alley[9] puts it, we each employ the equivalent of 100 serfs, each of whom consumes his or her 100 watts and produces the concomitant waste—at roughly 1.5 pounds per day, that's 55,000 pounds per year. If you want to start digging the latrine, it needs to be 10 feet by 10 feet by 9 feet deep—per person per year.

In fact, Alley notes that roughly 85 percent of these "serfs" come from fossil fuels that were created over a few hundred million years of the planet's evolution and that we are consuming over a few hundred years, i.e., a million times faster than Nature produced them. Contrary to various alarmist predictions, there is plenty of fossil fuel left to last for centuries at rates of consumption even greater than today's. But we cannot escape the waste; its impact is profound. Indeed, geologists, whose job it is to keep track of such things, have noted the recent transition to a new geologic era, the Anthropocene, in which the Earth's geological, chemical, and biological systems are dominated by one species: *Homo sapiens*.

Poe's wood nymphs will not help us escape from this situation, nor will Diana ride to the rescue. Likewise, an abundance of "clean" natural gas will not halt global warming, homeopathy will not cure cancer, and abjuring vaccines will not eliminate autism. No aliens, hiding in a comet's tail, will swoop down to save us. Only human imagination and ingenuity, channeled by the systematic and skeptical curiosity we call science, provide a viable path forward for our species.

This book is meant to be both a warning and a celebration. First, the warning: with mass media, entire political classes, and a general public largely ignorant of, or hostile to, science, our civilization cannot survive in its current form. This is not mere opinion but a statement of fact. How our civilization will change is a matter of speculation that I will largely eschew, but most of the plausible scenarios I can see unfolding in the near future are not pretty.

As important, however, is the celebration: the following chapters are written to share and, yes, to celebrate the illuminating habits of mind that characterize the enterprise we call science. This enterprise melds the marvelous natural curiosity of a five-year-old with the impressive reasoning power of the adult mind. It is a highly social enterprise, a highly creative enterprise and, William Blake notwithstanding, a life-affirming enterprise. Adopting these habits of mind opens up worlds both unseen and unseeable to understanding. It allows us to read the history of the deep past and to predict the future. It provides us with context: our "pale blue dot" is but one of eight planets and a few dozen moons orbiting one of a hundred billion stars (many of which, we now know, also have planets) that make up the Milky Way, one of a hundred billion galaxies in our visible corner of the universe. That, far more than a "rainbow in heaven," is awe-full—it inspires awe.

Perhaps most importantly, science allows us to transcend our evolutionary heritage by recognizing that the quick responses and "common sense" produced by a brain shaped over hundreds of millennia of hominid evolution in small hunter–gatherer groups are highly inappropriate on a human-dominated planet. Albert Einstein reportedly said this of common sense: "It is that layer of prejudices laid down in the mind prior to the age of eighteen." Either we abandon those prejudices or the Anthropocene will end nearly as abruptly as did the dinosaur's Cretaceous era—and without the assistance of an asteroid.

I should be clear. Life on Earth will not be extinguished by our folly. It will change and adapt to the new conditions we create, as it has changed innumerable times over the 3.8 billion years since it emerged in sufficient profusion to leave a record. The Earth will continue to orbit the Sun, and the Moon will still roll around in its space-time dimple for another six billion years or so until the Sun begins to die. It is only if we have affection for our kind, the species we named *Homo sapiens*, that we must adopt scientific habits of mind and relegate the wood nymphs to their proper place—as entertaining fantasies produced by our fertile imaginations but wholly irrelevant to the operation of the universe.

Homo sapiens means "wise man." Our behavior in the coming century will determine whether this is apt nomenclature.

* * *

By the scheduled end of my class, eighty minutes after having left the lecture hall, we made it to 86th Street. We had discussed why the Hudson

River is actually a fjord and had adduced not only the presence of the glacier that helped carve it but also the direction from which it ground its way forward by studying the striations in the Manhattan schist outcrops of the park. Later, plotting this direction on a map of the area, it was obvious that the lines of motion were precisely perpendicular to the orientation of Long Island, tying in this 100-mile-long mound of glacial rubble directly to the last great force that shaped the New York City landscape before men with pile drivers arrived a century ago. We discussed how the oldest trees in the park held a record of the temperature and humidity over the last century and a half in the isotopic composition of their annual rings and why the crocus blossoms look very different to us than to a bee whose sensitivity to ultraviolet light gives it a more richly variegated view,[10] as well as helping it to navigate by the polarization angle of the scattered sunlight that comprises the blue sky.

Before I let them go, I made my pitch for science—not as a course one has to take to fulfill a distribution requirement but as a way of greatly enriching one's view of the world. Seeing a flower as a carefully programmed array of molecules whose subtle structural differences mean that they absorb and reflect different wavelengths of light—including some only their pollinators can detect—deepens the aesthetic experience of seeing the flower. And understanding that "seeing" is a passive activity in which packets of electromagnetic energy bounce off a petal in all directions, with only a tiny fraction heading in a perfectly straight line for the little black dot in the middle of your eye. There, they pass through, focused by a transparent lens, and are absorbed by one of three special molecules tuned to red, green, and blue light such that they can induce a set of electrochemical signals that travel at ~10 meters per second along the optic nerve to the primary visual cortex at the back of your head and thence on to a dozen other cortical substations to create the vision of a crocus blossom (whose name—if you know it—can be dredged up from the brain's word storage area)—that's all pretty magical, too.

I'm sure not all of them bought it . . . my hope is some did. And we did get back to stellar nucleosynthesis in the next class.

2
WHAT IS SCIENCE?

A s I said in the first chapter, my goals for this book are both to deliver a warning and to share a celebration. Since celebrations are more fun, that will be my primary focus—to delineate some of the habits of mind that are the hallmarks of science and to provide examples of their application from the mundane to the sublime. In the process, I hope to illuminate a path to a richer and more rational view of the world. Simultaneously, I will strive to show how this path leads those who adopt it away from the unquestioning acceptance of misinformation as fact, falsehoods as policy, and myths as the basis for the future of civilization.

Our first task then is to be clear about what science is—and what it isn't. I must begin with a disclaimer: not all of my scientist colleagues will agree with everything I have to say on this subject. We came to science via different routes and, once arrived, have pursued divergent paths. Notwithstanding the sterile, six-step "scientific method" you were asked to memorize in junior high school, there are many ways to approach science, and most are considerably messier than the "method" lets on. There are many distinct scientific disciplines, each with its own cultural traditions and idiosyncratic conventions. And, of course, there are literally millions of individual scientists who, contrary to popular belief, do not all wear lab coats, thick glasses, and pocket protectors; indeed, I have never been in a lab coat, I wear drug store reading glasses, and I don't own a single pocket protector (although I will confess to both having a slide rule in my bottom desk drawer and long curly hair). What I provide here then is a personal view derived from four decades of doing science, watching other scientists, and reading the work of those who have thought about these matters more deeply than I have.

One thing all true scientists have in common is curiosity—we've each retained those insistent why, what, and where questions from childhood. In addition, we all adopt Nature as the ultimate arbiter of our models and theories, and we all press onward to refine our comprehension of Nature.

However, Truth—with a capital "T"—is not a matter of great interest. Truth is a human invention; Nature is the primary object of my concern. My finite mind can comprehend only a tiny fraction of Nature. However, that fraction is vastly larger than that comprehended by Cro-Magnon man, by the classical Greeks, or by Renaissance polymaths, not because I am smarter than these predecessors (in fact, average human brain size has contracted almost 10 percent since Cro-Magnon times[1]), but because of the enterprise we call science. This enterprise has created tools to extend the reach of our senses into realms far too small, yet too large to be viscerally experienced. It has collected far more data than any single individual could ever amass. It has systematized these data into models and synthesized the models into theories. It has created a body of knowledge (facts, models, and theories) and has refined a methodology for increasing that knowledge. This is science.

I no longer need to wait, as my hunter–gatherer ancestors did, for the Moon to rise each evening to see what shape it will be (or, indeed, to see whether it will rise at all). I have a mathematical model that predicts its shape and motions with great accuracy.

I don't need to speculate on where the medieval doctor should make an incision to bleed out the foul humors that sicken me with fever—I have an understanding of the microscopic creatures that cause the fever and the chemical means to defeat many of them.

I need not watch the clouds and wind incessantly to discern whether I must find shelter from an oncoming storm—I have orbiting satellites that keep a vigil on each storm's progress and computer models that predict several days in advance how the storm will move.

This does not, however, mean I have found the Truth—about the weather, about disease, or even about the Moon's orbit. I do have knowledge, and that makes life easier than in caveman times, but "Truth"?

Mathematicians are the ones who can prove things to be true. They have constructed axioms that define a system of reasoning within which it is possible to prove rigorously that an assertion (often called a theorem) is the Truth, with a fully justifiable capital "T." Mathematics, like the concept of Truth, is a creation of the human mind. We have defined all the terms, specified the rules, and so can, within this closed system that we alone control, define Truth.[2]

Scientists cannot prove things to be true. Indeed, scientists spend much of their time proving things—their models, mostly—false. Before you conclude this to be a dreary pursuit, look around you. Heinrich Hertz proved false the notion that light waves were restricted to the range of wavelengths our eyes can see—your cell phone follows from this falsification. Louis Pasteur disproved the theory of spontaneous generation—you are most likely alive to read this because he showed that such a common sense notion was wrong. Count Rumford disproved the fluid theory of heat—your last plane ride was made possible by his trashing of that intuitive idea.

The universe exists independently of us and of our constructions such as Truth. Indeed, it existed nearly 13.8 billion years before human consciousness emerged, and it will be around a whole lot longer than that after we are gone. The mathematicians' Truth will still be true 1,000 years, or 10 billion years, from now. Our scientific understanding of the world, by contrast, will not be the same next year. We will have shown a few of our current ideas to be false, and that will be regarded as progress. Rather than a search for Truth then, I define science as a system designed to search for falsifiable models of Nature.

FALSIFIABLITY

Nearly a century ago, the then-young philosopher of science Karl Popper found himself in the heady intellectual ferment of post-war Vienna where exciting new ideas were emerging from disparate fields: Einstein's relativity, Freud's psychoanalysis, and Marx's dialectical materialism. Popper recognized a fundamental distinction that separated Einstein's theory from those of Freud and Marx: relativity was falsifiable. It was cast in such a way that it made precise, testable predictions as to how nature should behave, and a single observation inconsistent with those predictions would destroy it. The same could not be said for psychoanalysis or Marx's theory of history (then or now).

Popper later summarized his conclusions about theories as follows: "One can sum up all this by saying that *the criterion of the scientific status of a theory is its falsifiability, or refutability, or testability*" [emphasis in original].[3] In reflecting upon his characterization of science as defined by the falsifiability of its models, Popper noted that

> what worried me was neither the problem of truth . . . nor the problem of exactness or measurability. It was rather that I felt that these other . . . theories (of Freud and Marx), though posing as sciences, had in fact more in common with primitive myths than with science; that they resembled astrology rather than astronomy.

He went on to acknowledge that myths can become testable, that many, if not most scientific theories, "originate from myths, and that a myth may contain important anticipations of scientific theories." Myths are, like science, the product of human imagination, but they do not enter the realm of science until they are cast in falsifiable forms. And then, once so transformed, they contribute to understanding.

Philosophers and sociologists of science are, in general, not fond of Popper's strict falsifiability criterion. The philosopher Susan Haack lists "a preoccupation with demarcation" (dividing science from other forms of inquiry) as one of her "six signs of scientism," which she decries as an "overly enthusiastic and uncritically deferential attitude toward science."[4] In his later writing, Popper himself softened the edges of his line of demarcation. For example, his categorization of the theory of evolution as a "metaphysical research program"[5] (and thus, not science) changed into an acceptance of evolution as part of a broader definition of science that encompassed any empirical way of knowing the world.

I, however, still find falsifiability a useful test, as the introductory anecdote in chapter 9 will make clear. Today, we can observe evolution directly in a petri dish yeast culture—or in the changing beak sizes of Galápagos finches in response to changing climatic conditions.[6] It is indeed a theory that makes falsifiable predictions, and scientists, using techniques as varied as those of the field ornithologist and the cell biologist, are testing evolution's predictions. Since it has yet to be found wanting, it remains a part of science.

It is, of course, inappropriate to be "uncritically deferential" to the practice of science or of any result it produces—indeed, it would be unscientific to be so, because skepticism is a key feature of a scientific mind. Furthermore, science is a human activity and thus includes among its practitioners its share of the mistake-prone, the incompetent, and the fraudulent. But science includes powerful antidotes to both error and deceit embedded in its social structure as well as in its prostration before Nature. If a theory diverges from Nature in even a single instance, it cannot be rescued by power or money, by bombast or divine authority. It must be modified or die.

FROM MYTHS TO SCIENCE

In ancient times, it was reasonable to ascribe the wind to an exhalation of Aeolus that varied with his mood or the commands of the gods[7]— reasonable in the sense that it was a self-consistent picture consonant with the social imagery and understanding of its time. But once one recognizes that air is made up of molecules of matter, once one invents the notion of pressure gradients and links them to temperature—once one has a falsifiable model of how and why the air should move—this myth is no longer required, and wind enters the realm of science.[8]

Our senses gather input from the world around us. Our brains record these inputs, search for patterns among them, and synthesize past and present experiences into tentative models of the reality in which we are embedded. At first, these models are unconstrained by the experiences of others or any awareness of the ways in which the world actually works (picture a toddler seeing his mother's smile and running toward it despite the fact she is standing on a flight of stairs several steps below him). Gradually, we recognize patterns of constraints—one always falls down the stairs, never up—and begin to incorporate these patterns into our worldview. Later still, we can share the experiences others have of the world and enlarge our understanding beyond our immediate sensations (the fact that lions eat little children can be understood without ever encountering a lion). We come to recognize the value of prediction to survival (those who don't are lost from the gene pool). Ultimately, we build up a collection of mental models that help us to navigate the world.

Science systematizes this sequence. Observations are made consciously, often quantitatively, and are carefully recorded. The accumulating experience of generations is incorporated into the record. Models are consciously built, constructed to encourage falsifiable tests, and then vetted for consistency with all that is known. Experiments are performed and further observations collected. Inconsistency between a model and the experimental results leads to the model's death; consistency is but a temporary reprieve until more demanding tests are made. Models are collected, linked, and forged into theories. Scientific understanding advances.

I use "understanding" here not in the sense of "perceiving the meaning of" but to describe an appreciation for the underlying cause of an observed phenomenon—and being able to use that appreciation to predict the future. Scientific understanding must always be tentative, but this does not make it un-useful. Indeed, it is easy to argue that science has led to more useful things than any other invention of humankind. So again, my definition: science is the process by which we systematically search for falsifiable models of Nature.

THE ATTRIBUTES OF SCIENCE

In working toward this definition of science, I have used words such as "data" and "models," "intuitive" and "messy." Some of these require more precise definitions themselves, as they are part of the essence of science. Others need a social context, both as aspects of science, and in contradistinction to it. In a trenchant essay entitled "Science in General Education,"[9] Andrew Read, a professor of biology and entomology at Penn State, has listed ten attributes of science that, although not all widely appreciated, both describe it well and provide a rationale for why it is an essential mode of thought for navigating the modern world. Since all of these attributes will emerge along the way, I list them here, and describe briefly how each is relevant to the acquisition of the scientific habits of mind I will describe:

1. Science works.
2. Science is extraordinarily effective at rooting out rubbish.
3. Science is antiauthoritarian.

4. Science struggles to deliver certainty.
5. The scientific process is messy.
6. Not all data are equal.
7. Science can explain the supernatural.
8. Science generates wonder and awe.
9. Science is counterintuitive.
10. Science is civilizing.

1. **Science works.** Toddlers *do* always fall down stairs, never up, and the Moon appears in a predictable pattern each night. Both of these happen because of a feature of the universe we call gravity. For 250 years we had a fine model for gravity that allowed us to predict precisely the location of yet undiscovered planets[10] and to construct elevators to avoid the stairs. That model, it turns out, failed to predict the precise position of Mercury as it orbited the Sun—over the course of a century, its position was off by the width of a human hair held at arm's length. But science is not like horseshoes—being close is not good enough. Newton's model of gravity was falsifiable—and the tiny error in Mercury's position presented a serious challenge.

We now have a new model of gravity—general relativity (which is a century old this year)—that gets Mercury's position spot-on and passes every other test we have devised. We are quite sure it is wrong too because it is inconsistent with our current model for matter at the subatomic scale, but it allows your GPS to work remarkably well, so we'll stick with it for a while longer. Remember, neither Newton's universal law of gravitation nor Einstein's general relativity are to be equated with "Truth"—they are models we have that describe how the universe behaves—and they work very well indeed.

Ignorance of, and antipathy toward, science is not at all uncommon today, but when not focused on their favorite myths, most people, in fact, live their lives assuming that science works. I have a colleague who has more than once become irritated with her postmodernist faculty colleagues in other fields—advocates of personally constructed realities and mostly hostile to the "hegemonic patriarchy" of science. Her response is usually to tell them to write about it on a postmodern laptop while flying on a postmodern airplane to their next conference. Not a single one has taken her up on it yet.

2. **Science is extraordinarily effective at rooting out rubbish.** As this will be a recurrent theme throughout the book, I will not elaborate extensively on this claim here. But, as noted in the preface and discussed further in chapters 11 and 12, there is an enormous amount of "rubbish" (I generally use the less emotional term "misinformation") abroad in the land today, and by far the most effective means of sorting information from misinformation is by applying the methods of science.

3. **Science is antiauthoritarian.** This statement might come as a surprise to some readers. Indeed, we frequently hear references to scientific "authority," and anyone who has visited a large, structurally hierarchical research lab would be forgiven for inferring that an authoritarian regime is in place. But the essence of science is that the only authority is Nature itself. When making a scientific argument, appealing to an authority other than Nature is forbidden. One does, of course, cite previous work, accept (preferably after testing them) data collected by others, and usually defer to those with greater knowledge or experience in a given field. But if an undergraduate research assistant comes up with a result contrary to the theory espoused by the lab's director, it must be granted a hearing and, if confirmed, supplant the theory in question.

Does this happen in practice? Perhaps not frequently, although there are many instances in the history of science of just such an occurrence. The point here, however, is to describe the essential attributes of science, and antiauthoritarianism is certainly one of them. The mathematician and historian of science Jacob Bronowski put it succinctly:

> The society of scientists . . . has to solve the problem of every society, which is to find a compromise between the individual and the group. It must encourage the single scientist to be independent, and the body of scientists to be tolerant. From these basic conditions, which form the prime values, there follows step-by-step a range of values: dissent,

freedom of thought and speech, justice, honor, human dignity, and self-respect.[11]

These are, one and all, antiauthoritarian values.

4. **Science struggles to deliver certainty.** As argued previously, science strives to provide an accurate model of the world but, by my definition of its aim, does not deliver certainty—the "Truth." It can sometimes seem that a scientific prediction is certain enough to be accepted as Truth. The Sun will rise in the east tomorrow morning; you will probably find few people in full possession of their senses to bet against this proposition. But it is not a scientific statement to say that this is a certainty. Science is about building falsifiable models and, without the ongoing struggle to find their faults, we cease to do science.

Science's progress over the past 400 years has been spectacular—we can now predict the position of Mercury in its orbit to much better than the width of a human hair over 100 years (the error that spelled the demise of Newton's theory). Yet we know (1) that the model used to make this calculation is almost certainly incomplete if not entirely wrong and (2) that planetary orbits are chaotic and therefore not predictable over hundreds of millions of years; thus, we have not produced certainty. That's just the way it is. Science has enormous powers of prediction, just not absolute power.

5. **The scientific process is messy.** This is why I dismissed the steps of the junior high version of the scientific method as not worthy of recitation. First, there is the problem of making good observations. We are always limited by our senses or by our instruments, by the influence of extraneous factors, by our prejudices and expectations, and by the noise inherent in all physical processes. Data are never perfect. Then there are the limitations of the human mind, shaped by past experience, primed to see patterns where none exist, constrained by limited computational capacity, and wary of the unknown. On top of this are the imperfections of any human social activity (and, again contrary to popular belief, science is a highly social enterprise): unwarranted egos, competition for funding, self-delusion, political machinations, etc.

Messy, but with very powerful self-correcting mechanisms in place: peer review of published work, the requirement of sharing data and methods to facilitate reproducibility, and, ultimately, the authority of Nature, which, as physicist Richard Feynman once said, "cannot be fooled."

6. **Not all data are equal.** All data are imperfect, and not all data are equal. An anecdote cannot substitute for a carefully designed, double-blind study; as an oft-repeated phrase has it, the plural of anecdote is not data. Data collected with primitive instruments are inferior to those taken with the best available techniques. Data taken with the fewest confounding influences possible are superior to those derived from complicated situations. And data must come with their uncertainties—quantitative estimates of how closely our measurements have come to representing the quantities of interest. Data without their associated uncertainties are useless (see chapter 7).

7. **Science can explain the supernatural.** The myth of Aeolus is relevant here. Supernatural forces mean forces controlled by something/someone apart from Nature. Absent models for the workings of Nature, it is unsurprising that the creative powers of the human mind invented stories to account for the unexplainable—from the origin of the winds to the origin of the Earth itself, from foul humors that cause disease to the fact you had a bad day yesterday (the stars were misaligned), etc. Science cannot fully explain all of existence, so supernatural beliefs continue to flourish. But we've done a pretty good job on the wind, which means there are very few worshippers of Aeolus still around. We've made progress on disease as well, so blood-letting is way down. On the origin of Earth there is still a lot of competition, although more on that in chapter 9. As for your bad day, statistical fluctuations (you have some good days and some bad days) rather than causative factors might be worthy of consideration.

It is worth noting here what the essayist Philip Slater wisely observed: "Man is unique among animals in his practiced ability to know things that are not so."

Think about it. How many monkeys do you suppose believe that big coconuts fall faster than small coconuts? The educated elite of Western civilization were convinced of this Aristotelian notion for roughly 2,000 years. How many dogs do you think reason: "Oh boy, Venus is in the sixth house so it's going to be the green Liv-a-Snaps tonight!" Only humans conceive of supernatural forces. Only science can liberate us from this conceit.

8. **Science generates wonder and awe.** This is my theme in chapter 1 and will remain a theme throughout, so further examples are not needed here.

9. **Science is counterintuitive.** This may bring a smile to the lips of those of you who struggled with science in school, but it is meant in a more fundamental sense. The evolutionary line that led to humans diverged from chimpanzees roughly six million years ago. Our brain has been evolving continuously since that time. It has mainly gotten bigger and better at allowing us to find food, to avoid predators, and to reproduce. These activities require quick reactions, an ability to see patterns, and a whole array of autonomic responses in the brain about which we are barely conscious. By trial and error, through natural selection, our brains are highly tuned to accomplish these survival tasks.

More than 99.9 percent of the time this brain was evolving, however, we had no writing and no numbers. More than 99.993 percent of that time we had no science. Evolution is a slow process. Galileo's daughter's brain was not systematically different than his, and neither are yours or mine. The cognitive scientist Keith Stanovich illustrates this beautifully in his book *The Robot's Rebellion*.[12] Our brains are programmed to make sense of the world as it was 10,000 (or 1 million) years ago, making science—a powerful yet very recent invention—fundamentally counterintuitive. We will encounter a number of illustrations of this as we proceed.

I saw a bumper sticker recently that sums it up nicely: "Don't believe everything you think." Science is a way of deciding which of the things you think are worth keeping.

10. **Science is civilizing.** Although Blake and Poe may have profoundly disagreed, one can do little better than harken back to Bronowski's declaration that science encourages "dissent, freedom of thought and speech, justice, honor,

human dignity, and self-respect" —civilizing qualities all. It replaces what Carl Sagan called a "demon-haunted world"[13] with a world in which an understanding of nature fundamentally changes the human condition.

My dictionary defines "civilize" as "to bring to a stage of social, cultural, and moral development *considered* more *advanced*" [emphasis added]. In light of this problematic description ("considered" by whom?; "more advanced" in what sense?), I would step back and say that science is a civilizing force—quite possibly in light of human history and prehistory, an essential adjunct to the civilizing process—but it is not, at least yet, ready to take on this task alone. In this book I will focus on describing the habits of mind that science contributes to this project we call humanity.

* * *

These ten characteristics of science stem from a mindset built upon a specific set of habits that are employed when interacting with the world. In each of the following seven chapters, I highlight one of these habits of mind. I begin each chapter with an anecdote from my experience and then systematically explore applications of the featured habit from the quotidian to the esoteric—from personal decisions about medical care to what it was like during the first nanoseconds of the universe. For those who wish to practice scientific habits of mind, problems are provided in the appendix.

It is my contention that these habits are liberating and empowering. They transform the mind from a passive receptacle for information and misinformation—fed in by those with titles, those with economic power, and those with a political agenda—to an active player who can challenge, dispute, analyze, and evaluate information and use it wisely to inform decisions. Cultivating these habits is the most efficacious approach to surviving the Misinformation Age—and for helping one's fellow citizens to survive it as well.

3
A SENSE OF SCALE

"10 a m. 16 Peck Slip."

E-mail sometimes reminds me of the clipped cadence of British officers in old war movies. In this case, the message *was* a call to arms of sorts. It came from Dr. Neil deGrasse Tyson, Director of the Hayden Planetarium, and was sent to inform me of the time and place of our appearance before the New York City Commission on Landmarks and Preservation. We wanted to knock down a protected building.

The massive granite facade of the American Museum of Natural History has presided over the Upper West Side of Manhattan for well over a century. The museum, its echoing halls lined with an unimaginable menagerie of stuffed birds and animals, ancient Pacific war canoes, glistening minerals, and, of course, dinosaurs, has introduced generations of New York City school children to science. Nestled behind its facade, on the north side of the building, was one of the museum's more recent additions—the (then) sixty-year-old Hayden Planetarium.

When the planetarium was built in 1935, the New York night sky had long been ablaze with the illumination required by the city that never sleeps. The Milky Way was already a candy bar, not a swath of opalescent light girding the night sky; at most a few dozen stars were visible on a clear night in Manhattan. Yet inside Hayden's dome ("the first example of sprayed concrete construction," the architectural historians solemnly informed us), the night sky came alive for thousands of school children as the stuffed birds and dinosaur skeletons never could. The graceful sweep of the silver projector in the center of the room could carry them through the four seasons in a minute, to the Southern Hemisphere and back in less, or just upstate to an Adirondack hamlet where the stars still lit the evening sky. I have yet to meet a New Yorker between the ages of nine and ninety who hadn't taken these magical journeys or weighed him or herself on Mars and Jupiter on the scales appropriately calibrated to the gravity of each planet at the planetarium.

And we wanted to knock it down.

The inexhaustible supply of would-be politicians, self-proclaimed activists, and genuinely concerned citizens who lived within thirty blocks of the museum guaranteed that our quest would be an interesting one. The political connections and academic credentials of the Commission's members required us to prepare a compelling case. We felt we were ready.

The architectural firm of Polshek and Associates had designed a breathtaking replacement for the old planetarium. The museum, under President Ellen Futter's energetic leadership, had come up with the money to finance the $210 million project. We were going to keep the Martian scales and the Art Deco doors, although the sprayed concrete would have to go. In its place, we would erect a structure in which the full span of space and time would come alive—a vantage point from which visitors could gain an objective perspective on their unique place in the universe.

Peck Slip. I had never heard of it, so I had to dig out my disintegrating street map of Manhattan from the bottom of the dresser drawer (this was before Google maps) and search for it. Way, way downtown, in the tangle of streets laid out in lower Manhattan when it was New Amsterdam, I found it, projecting into the East River. A quay, really, no more than 200 m long—no need to leave early to find number 16.

The building at 16 Peck Slip is "landmarked" itself. Once through the metal detectors (I guess some people get really passionate about old buildings) and up the stairs, I emerged into the hallway outside the Commission's imposing hearing room. Thirty-foot ceilings, a table constructed at the expense of a small forest—it was all on a grand scale, an appropriate venue for lofty civic principles and the sweep of history that gave context to the Commission's work.

Our presentation had been carefully prepared and thoroughly rehearsed. President Futter delivered articulate and forceful testimony about the museum's vision of its role in the twenty-first century. The architects gave a sumptuously illustrated presentation of the new design, as well as a carefully reasoned explanation of why the current facility was hopelessly inadequate. Tyson eloquently described both his own childhood experiences at Hayden and his vision for its future with an ineluctable enthusiasm undiminished since his first visit. I was there to weigh in as chair of the Department of Astronomy at Columbia University with a dispassionate academic endorsement. Because academics feel most comfortable giving lectures, I delivered a lecture on the scale of space and time.

The centerpiece of the new planetarium's design was to be a smooth, eighty-seven-foot-diameter ball suspended inside a ten-story glass cube. The visual effect would be magical, but it was the possibilities for conveying a visceral sense of the scale of the universe that made the design irresistible. Imagine the nine-story sphere as the Sun. If constructed to scale, the Earth would be a ten-inch pale blue and green ball. We could hang it from the ceiling—London and Dakar would be indistinguishable to the unaided eye—"[a] very fit consideration," in the words of the great seventeenth-century Dutch polymath Christiaan Huygens, "and matter for reflection, for those Kings and Princes who sacrifice the lives of so many people only to flatter their ambition in being Masters of some pitiful corner of this small spot."[1]

Now, the difference in diameter between the Earth and Sun is only a factor of a hundred; the Earth is just under 8,000 miles across, whereas the Sun is a little over 800,000 miles in diameter. The Earth's distance from the Sun, however, is more than a hundred times the Sun's diameter. If we positioned the Earth to scale, then our ten-inch globe would end up orbiting the planetarium Sun nearly two miles away, through Rockefeller Center. On this scale, the asteroid belt, that debris pile of early solar system material that never managed to form a planet, would be located around City Hall. I suggested, following Huygens, that locating it at the seat of city government might serve as a gentle reminder to that building's occupants of their importance from the perspective of the universe. Fortunately, the commissioners (mayoral appointees all) had a sense of humor.

In this New York-sized solar system, Mars would be a five-inch pink ball passing through 25th St. and 133rd St. in its orbit. Neptune, the most distant planet from the Sun, would be a glowing green sphere three feet across about 53 miles up the New York State Thruway with the following inscription: "You are now 2.8 billion miles from the Sun and 53 miles from Hayden Planetarium." For the nearest star, we did have a problem. Another eighty-seven-foot sphere in Perth, Australia (the closest city to New York's antipodes[2]), would be forty-two times too close. Indeed, even the Moon is a factor of two too close for siting—to scale—our nearest stellar neighbor.[3]

But it was not just large scales the new planetarium could illustrate. If the sphere represented an atom of carbon in your eyelash, that atom's nucleus would be the size of a small grain of sand. If the eyelash were made of atoms this large, it would be more than twice the width of Asia and stretch ten times the distance to the Moon.

The commissioners got the point, signed the epitaph for the historical sprayed concrete, and cleared the way for a major new addition to the cultural highlights of New York City: The Fredrick Phineas and Sandra Priest Rose Center for Earth and Space. At night, the suspended sphere is indeed magical; by day, it serves as a window through which visitors can begin to grasp the scale of the universe.

ARE HUMANS BIG OR SMALL?

Both of course. Over the last hundred years, science has developed a perspective on our place in the universe by observing particles both inside the atomic nucleus and billions of light years away in space. In doing so, we have uncovered the basic forces that govern the cosmos, woven them into the fabric of space and time itself, and reconstructed the history of the universe and all it contains.

We begin then by exploring the range of physical scales that make up the domain of science. This range is enormous and leads to two immediate problems: (1) the numbers used to describe the range are unimaginably large (and small), and (2) the distances lie far outside the realm of our experience and thus, by definition, don't comport with our "common sense." The goal of this chapter is to help you transcend the limitations of your senses and experience and to look objectively at the full range of scales in space and time that shapes our current understanding of the physical universe.

THE LIMITS OF PERCEPTION: THE SCALE OF SPACE

First, let us examine the origin of our perceptions of small and large. What are the limits of the human senses? What is, for example, the smallest distance you can directly perceive?

The two senses one might think of as relevant to this problem are touch and sight (hearing, smell, and taste don't sample distances in space directly). Let's start with touch.

Have a friend (and make sure it is a friend) conduct the following experiment. Get two very sharp pencils or a divider (the device used to determine distances on maps). Put a blindfold on and have your accomplice hold the two points close together (less than a quarter of an inch apart) and touch you gently—and absolutely simultaneously—on your shoulder blade. Can you tell if it is one prick or two?

Conduct a series of tests to determine the minimum distance at which you first feel two separate stimuli. Have your friend vary the spacing between the points until you can correctly identify the two pricks as separate stimuli. To test the authenticity of the answers, the pricker should randomly alternate between one prick and two.

Even though the nerve endings being stimulated by the prick are very tiny (much too small to see with the naked eye), the elaborate process by which contact between the pencil's carbon atoms and your skin molecules is translated into a sense of being "touched" has a rather limited sensitivity. On your shoulder, where nerve endings are relatively scarce and the portion of the brain devoted to recording their activity is relatively small, distances less than roughly half an inch apart cannot be resolved. Other "more sensitive" areas of your skin might do better. Repeat the experiment (again blindfolded) on the tip of your index finger. What is the minimum perceptible distance here?

Take off the blindfold when you reach the separation at which you can no longer feel two pricks. Can your eye distinguish the two distinct points that touch failed to record?

If the answer is yes (and if it isn't, you are badly in need of glasses), perhaps the eye is a better tool for probing small distances. Let's test it.

Pluck a hair from your head. Stretch it between your fingers and hold it taut at arm's length. The width of the hair is still visible. Now have someone hold it up twice as far away. This is equivalent to looking at something half the width of a hair strand (an alternative approach would be to slice the hair in half lengthwise, but just doubling the distance is easier). Can you still see it? Double the distance again. Now, I suspect, your eyes can no longer discern it. This width, roughly one-quarter that of a strand of hair, is "too small" for you to see.

Why? (There's my favorite word again.) What sets this limit?

Your vision is limited by two numbers that, at first, may seem completely unrelated to the issue at hand. These numbers are the surface temperature of the Sun and the strength of gravity on the surface of the Earth. The former determines the colors, or wavelengths, of light emitted by the Sun, and the latter both (1) sets the composition of the Earth's atmosphere (and thus determines which wavelengths of sunlight get through to the ground) and (2) establishes a limit on the practical size of living things.

The surface temperature of the Sun is 5,780 Kelvins.[4] The Sun–atmosphere combination leads to an abundance of light centered at a wavelength of 500 nanometers (1 nanometer = 10^{-9} meters)—what we normally call "yellow" light. It is not a miraculous accident that the peak sensitivity of our eyes occurs at this same wavelength. Four hundred million years of evolution has carefully tuned our natural radiation detectors so that they are most sensitive to the wavelengths where the most light exists—the better to see tasty morsels of food and to avoid becoming such a morsel oneself. Of the seventy octaves of radiation the universe sends our way, we "see" only one—from about 360 to 720 nm. Our eyes, then, provide a sadly impoverished view of the cosmos. Our ears, at least, can hear ten octaves of sound waves—imagine how awful Beethoven's Ninth Symphony would sound if you could only hear it one octave at a time (or, if you can't imagine that, experience it by visiting my website[5]).

Given this narrow range of wavelength sensitivity, the other quantity that sets the limit on our ability to see small things is the size of our eyes—or, more precisely, the diameter of the small black spot in the center of the eye through which the light enters, the pupil. Your pupil changes size somewhat depending on the brightness of your surroundings, but the average diameter is about 0.5 cm. The smallest angle we can observe directly is given by the ratio of the wavelength of light to the diameter of the detector, or, in this case, wavelength/diameter = 500×10^{-9} m/5×10^{-3} m = 1×10^{-4}, where this value is the angle measured in radians.[6] The angular diameter of an object is represented as S in figure 3.1, and, for small angles (less than a few degrees), it is well approximated by the relation

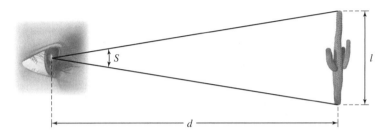

FIGURE 3.1 ANGULAR DIAMETER DEFINED

How we define the angular diameter S (or the angle subtended by) an object of size l at a distance d.

$S = l/d$, or angular diameter (again in radians) = physical diameter/distance. The hair, held at arm's length, thus has an angular diameter of 0.1 mm × $(1 \text{ m}/10^3 \text{ mm})/1 \text{ m} = 1 \times 10^{-4}$,[7] which we found previously is, in fact, about the resolving power of your eye. If you double or triple the distance, your eye can no longer see the hair at all.

Thus, it seems that the practical limit of human vision is a distance just a few times smaller than the width of a human hair—a distance of about 0.1 mm. Does that make a hair the smallest thing that exists? For most of humanity's journey on Earth, the answer would be yes. It takes a philosopher to speculate on the nature of things insensible to human vision—or a scientist to develop techniques to measure much smaller scales.

Now let's consider the largest distance you can directly perceive. You can, of course, fly from New York to California, or even to Tokyo nonstop, and by the time you get off the plane, it feels like a very long way. But you have no direct perception of the distance. Indeed, sitting in a plane on the ground waiting for the de-icing machine for a few hours is nearly as uncomfortable, and you haven't gone anywhere. Trains and cars do not help much, even though the latter allow you more direct access to the instruments that indicate how fast you are traveling and how far you have gone. But reading the odometer conveys a sense of distance only indirectly. For a direct sense of distance covered, only your legs will do.

What is the longest distance you can imagine walking? Take a long weekend and hike the Appalachian Trail (actually doing this is not required, but it might be fun). If you are in good shape and don't sleep too much, you might make 100 km. That seems like a pretty good upper limit to the distance you are ever likely to experience directly.

Now, let's compare the small and large limits of your distance perception. If I wanted to cover the trail for 100 km with individual human hairs laid side by side across it, how many would I need?

$$(1 \text{ hair width}/0.1 \text{ mm}) \times (10^3 \text{ mm/m}) \times 100 \text{ km} \times (10^3 \text{ m/km}) = 10^9 \text{ hairs}$$

The answer is one billion. That is, the ratio of the smallest distance you can directly perceive to the largest distance you might directly experience is a factor of a billion.

One billion is a large number. If you tried to count the one billion hairs on the trail, starting tomorrow morning, and counted 1, 2, 3, 4, 5 . . . at the rate of 20 numbers every 5 seconds, that would be 240 per minute, or about 100,000 if you did it 8 hours a day. You'd get to a billion (if you never took a weekend off) in 10,000 days or a little over 27 years. It takes a lot of hairs to cover a 100-km-long trail.

BEYOND DIRECT EXPERIENCE

Today we explore the natural world on scales far beyond what our senses can detect directly. What is a hair made of? If we peer down to length scales a billion times smaller than a hair's width, we are still on a scale that is too large for studying the constituents of the atomic nuclei that anchor the hair's atoms. And if we expand our view by a factor of a billion beyond our arduous hike, we will not have reached much beyond the confines of our solar system. Another factor of a billion beyond that and we are still trolling around in the local universe.

The full range of scales that science explores—from the quark–gluon interaction inside a proton (or the first moments of the universe after the Big Bang) to the edge of the observable cosmos—is about 100,000 billion billion billion billion or 100,000,000,000,000,000,000,000,000,000,000,000,000,000. Clearly, this is not a number that is easy to visualize. This is why we will use scientific notation[8] liberally in this book: the range is a factor of 10^{41}. That does not mean it is not a perfectly valid description of the realm of spatial scales we probe, but its size does mean that some analogies are needed to help us appreciate the scales of science. Here is an activity you can do in your local park that will help (at least a little).

MODELING THE UNIVERSE IN THE PARK

Get a tennis ball and head out to the park (as you can see, I like doing science in the park). This object is 6.5 cm (0.065 m) in diameter. And if it is one of those bright yellow ones, it makes a great Sun, so place it on a bench, or somewhere you can see it from a distance. Let's start by laying out the solar system.

This simply requires using ratios: the 0.065 m diameter of the ball is in the same proportion to the 1.39×10^9 m diameter of the actual Sun as, say, the scaled model of the Earth's diameter will be to the 1.27×10^7 m diameter of the real Earth or, in standard mathematical notation, 0.065 m/1.39×10^9 m = x m/1.27×10^7 m. Solving for x, we get that the scale model Earth should be 5.94×10^{-4} m or just about 0.6 mm in diameter—about the tiniest pebble you can find.

Where should we put the pebble? Since the scale remains the same—the ratio of the tennis ball to the Sun doesn't change—it is always going to be 0.065 m/ 1.39×10^9 m = 4.68×10^{-11}. Multiply this by any actual distance of interest, and you'll get the scale model distance. For the Earth's location in space, the actual distance is 1.50×10^{11} m, so the right place to put your little pebble is 7.0 m away (seven big strides if you are about six feet tall).

There's a nifty check you can do to see that this scale model is correct. Stand where the Earth is located and hold up the fingernail of your pinky and align it with the tennis ball. It should just about cover the tennis ball with no fuzz sticking out on either side. Now do the same thing by using your fingernail at arm's length, observing the real Sun from the real Earth (assuming it is clear out)—again, it should just cover the disk of the Sun (Don't look at the Sun without the fingernail obscuring it, however!). This means you must have the scale model correct, as the angular size of the "Sun" is the same in both cases (see the figure 3.1 to remind yourself what angular diameter is).

Depending on how ambitious you are and how many helpers you have, you could complete your model solar system using the planet diameters and distances from the Sun, as given in table 3.1.

TABLE 3.1 SIZES AND DISTANCES IN THE SOLAR SYSTEM

OBJECT	DIAMETER (KM)	SCALED SIZE (CM)	DISTANCE TO SUN (KM)	SCALED DISTANCE (M)
Sun	1,391,900	6.5	—	—
Mercury	4,866	0.02	57,950,000	2.7
Venus	12,106	0.06	108,110,000	5.0
Earth	12,742	0.06	149,570,000	7.0
Moon	3,475	0.016	384,403*	0.018*
Mars	6,760	0.03	227,840,000	10.6
Jupiter	142,984	0.67	778,140,000	36.3
Saturn	116,438	0.54	1,427,000,000	66.6
Uranus	46,940	0.22	2,870,300,000	134
Neptune	45,432	0.21	4,449,900,000	208

* Distance from Earth

Note that Neptune is a couple of football fields away, and you are going to have a tough time even seeing the tennis ball Sun without binoculars. The nearest star, Alpha Centauri,[9] is very similar in size and temperature (and therefore color) to the Sun, so it can conveniently be represented by another tennis ball. The problem is, this second tennis ball won't be in the park—even if your local park is the largest park in the world: the Northeast Greenland National Park (which is unlikely to be yours, since this 972,000-sq. km park has no permanent residents). Using our scaling ratio, the correct location for the ball representing Alpha Centauri is 3,860 km away, which if, like me, you were laying out your solar system in New York's Central Park, would place it just east of Phoenix, Arizona.

The only other star that would easily fit within the United States is the second closest star, Barnard's Star, which would be a marble-sized object near Juneau, Alaska. There are five other stars that are closer than Honolulu, but they lie in different directions so would end up bobbing in the ocean. In fact, if we used the entire surface of the globe to position the nearest stars, only fourteen systems would be found. Two of these are triple systems and four are binary pairs, making a total of twenty-two stars. Only two, Sirius A and Procyon A, are bigger than the Sun; on our scale, Sirius would be the size of a bocce ball and Procyon a duckpin bowling ball. Two of the other stars are white dwarfs, the corpses of dead stars that are only grains of sand like the Earth. The remaining seventeen (excepting Alpha Centauri, which is another tennis ball) are all marbles.

We stand then on our tiny pebble in the park looking out across the entire real Earth and see one bowling ball, one bocce ball, two tennis balls, seventeen marbles, and two other tiny pebbles. That's it. A few of these have even smaller pebbles orbiting them—even the nearest system appears to have such a planet. But with these few dozen objects spread out over the 510 million sq. km of Earth, it's a pretty empty place.

In addition to these twenty-two neighbors, there are one hundred billion more stars, scattered with similar spacing, that make up our little island galaxy, the Milky Way. So the Milky Way is pretty big. If the real Earth were now shrunk to the size of the tennis ball—with those twenty-two stars inside—the center of the galaxy would be 145 m away. We'll leave the question of the other hundred billion galaxies aside for now.

MODELING ATOMS IN THE PARK

Making models of scales too small to see is instructive as well. Use the same tennis ball to represent a carbon atom. Since a real carbon atom is about 7×10^{-11} m across, the scale factor we'll be using in this case is 0.065 m/7×10^{-11} m = 9.29×10^8. So if an eyelash were made of atoms the size of the tennis ball, it would be 10^{-4} m $\times 9.29 \times 10^8 = 9.29 \times 10^4$ m or 93 km (55 miles) wide—way bigger than the park. This might also help you imagine how many atoms there are in that eyelash—just think of a rectangular box of tennis balls 93-km wide, 93-km high, and stretching all the way to China. It would take about three hundred million trillion (or 3×10^{20}) tennis balls to fill the box.

Going to even smaller scales reveals a remarkable feature of the atomic world. In your junior high school science texts, atoms were represented as little solar systems with the nucleus replacing the Sun and electrons playing the role of the planets orbiting about it. The electrons are typically about four or five nuclear diameters away in those textbook illustrations or about a foot from the tennis ball. In fact, however, atoms are 99.999999999999 percent empty space. If you let the tennis ball act as a carbon nucleus, its six electrons (find a few more small pebbles) are orbiting in two distinct rings about 370 m (four football fields) and 1.3 km (0.8 miles) away. It is remarkable then that with all that empty space you can't simply pass your hand through this book, isn't it?

Although these park-sized models do not help a lot in visualizing the full range of scales we study (over forty-one factors of ten), they should provide some sense of the scale of things that lie far beyond our direct perception. But enough on the scales of space for now. Let us explore a scientific perspective on scales of time.

THE LIMITS OF PERCEPTION: SCALES OF TIME

We can begin again with the limitations our senses impose on the lengths of time we can directly perceive. The eye, for example, has a "flicker–fusion" threshold. That is, if a light is turned on and off slowly enough, one perceives it as a series of flashes, but if it is flipped on and off faster than this threshold, it appears as a continuous, steady glow. The flicker–fusion threshold varies with such parameters as the average brightness of the light, the blackness of the off periods, and where within your field of vision the light source is located. It also varies from individual to individual. Roughly, it is twenty-five times per second. One cannot then "see" anything happen on a timescale shorter than about 1/25th of a second (which is why movie and TV cameras take twenty-five frames per second to present a continuous moving image).

The ear has a similar limitation in distinguishing separate sounds—at roughly twenty ticks per second the sound merges to form a continuous hum. Indeed, a fundamental limitation of all our senses is set by the speed at which signals travel through our nervous system (between roughly 10 and 100 m per second) and the rate at which individual neurons respond (a few milliseconds). Thus, we cannot perceive directly any timescale shorter than about 0.02 seconds.

On long timescales, we have a very obvious limit—our lifetimes. The amount of Bach your mother played to you in the womb notwithstanding, your direct perception of a "long" time is the number of years you have been alive. You might be able to extrapolate a bit—to imagine what it must be like to have lived three or four times as long, but that's about it. You may have visited the California redwoods and been told they were 2,000 years old (or the Pyramids which are 4,500 years old), but you certainly do not have a visceral feeling for this longevity.

Again, science has greatly expanded the timescales we can explore.

Actual demonstrations such as the tennis ball in the park are more difficult (and/or tedious) in the temporal realm, so we will settle for some useful analogies. Although your lifetime is certainly an upper limit to the length of time you can directly perceive, and even that is a little hard to imagine—you cannot, I suspect, remember much from your first five years, for example. One year seems a comfortable scale to deal with. You can probably recall where you were last year at this time, and you can reconstruct the major events in the intervening months. Thus, we will take a year—one orbit of the Earth around the Sun (365.25 days or 3.16×10^7 seconds)—as our basic span of time. What are the temporal frontiers that science explores?

The upper limit of time for a cosmologist is the age of the universe. That the universe even has a finite age is a relatively new concept; from Aristotle to Einstein, the prevailing view was that the universe always had been and always would be. But in 1964, a single observation overthrew that orthodoxy. In the intervening five decades, we have measured the universe's age with increasing precision; we are now confident that we know the value to an accuracy of a quarter of one percent: 13.798 ± 0.037 billion years ($1.380 \pm 0.004 \times 10^{10}$ years).[10] We can reconceptualize the major events in the evolution of the universe, the Earth, and humankind by creating a comfortable scale such that 1 year = 13.8 billion years. On this scale, 437 years pass each second (so Sir Francis Drake, in his circumnavigation of the globe, sailed through the Straits of Magellan one second ago).

The universe, and time itself, began in the Big Bang. Let's take that as the moment on New Year's Eve when the big ball hits the bottom of One Times Square. Within one one-hundredth of a microsecond of midnight, all the protons and neutrons that constitute the nuclei of all matter, all of the hydrogen nuclei, and more than 80 percent of the helium in the universe today, were produced. Fifteen minutes later (just in time for the next bottle of champagne), the first atoms were formed as itinerant electrons joined with free-running protons and helium nuclei. The light from this epoch, 377,000 actual years after the beginning, is detectable today and provides some of the strongest evidence we have that we live in a universe of finite age.

Over the next few weeks, tiny fluctuations in the density of matter began to grow. Regions containing more gas (and dark matter) exerted a slightly stronger gravitational attraction, and gas from the surrounding area fell toward these regions, further enhancing their attraction and drawing in gas from even greater distances. In our neighborhood, these large and glowing clouds of gas began to fragment into stars in mid-January.

By late March the large spherical halo of these now-old stars saw a spinning disk of gas form in its mid-plane that, in turn, made more stars and formed the island universe we call the Milky Way. Billions of stars blazed forth, then died. Some of these produced elements such as carbon, oxygen, calcium, and iron in their nuclear furnaces while alive, then died in spectacular explosions that distributed these newly formed atoms throughout interstellar space, where they became available to form the next generation of stars.

Around 2:30 A.M. on September 2, a little clump of gas containing some of this freshly synthesized material began to collapse into a star— our star, the Sun. Within an hour and a half or so, the Earth formed from the detritus left over from the Sun's birth.

Within three weeks, complex molecules floating in the Earth's primordial ooze learned the trick of self-assembly and replication, and life emerged. But it was not until December 16 around 4 P.M. that a sudden (geologically speaking) blossoming of life forms occurred—the Cambrian explosion. Complex animals and flowering plants appeared in the oceans and on the continents. On Christmas Eve (early evening on December 24), land animals reached their pinnacle (in size at least) as the dinosaurs ruled a lush, tropical Earth. Then, just after 6:10 AM on December 29, disaster struck in the form of a 10-km-wide asteroid that hit the Yucatán Peninsula in Mexico with the force of thirty million nuclear warheads, igniting wildfires worldwide and then plunging the Earth into a cold darkness that lasted for months and extinguished over half of all species of plants and animals on the planet. Over the following two days, however, this ecological disaster led to the dominance of new plants and animals, including some cute furry ancestors of ours, the mammals.

And now it is New Year's Eve. At 10 PM, with the party in Times Square already several hours old, the first hominids appear on the plains of East Africa. Over the next two hours, they evolve from upright apes fighting jackals for scraps from the carcasses of dead antelopes to, at 11:59:30 PM, cavemen emerging from a very long winter as the last Ice Age recedes on the warming Earth. These last thirty seconds of the year, it turns out, bring the most stable period of climate the Earth has seen in "days." Agriculture, then civilization, are born. The ball has been dropping all this time in Times Square.

At 11:59:49.6, with less than ten and a half seconds left in the year, those "ancient" pyramids were constructed for the burial of Egyptian pharaohs. Five seconds later, the rival schools of Aristotle and Democritus debated what the structure of matter is like on a scale too small to see. At 11:59:59.55, less than half a second to midnight, the longest enduring democracy of the modern era is proclaimed in the Declaration of Independence. And, with only a tenth of a second to go (less than the blink of an eye), you were born.

Happy New Year, with best wishes for a new perspective on space and time.

INTERLUDE 1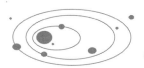

Numbers

As the preceding chapter suggests, numbers are an essential tool in the survival kit of an inquiring mind attempting to survive in the Misinformation Age. If you asked me the meaning of the word "number," I would have said something like "an abstract representation of things one has counted or measured in some way."

Google search results for "meaning of number"[1] may astound you (they did me). The first five hits are as follows:

1. "Spiritual Meaning of Numbers" (www.what's-your-sign.com)
2. "Meanings of Numbers from the Bible" (biblestudy.org/bibleref /meaning-of-numbers-in-bible/introduction.html from crystalinks.com)
3. "Angel Numbers" (sacredscribesangelnumbers.blogspot.com/p/index -numbers.html), which offers a "guide to repeating number sequences and their messages and meanings"
4. "Numbers and Their Meanings" (www.crystalinks.com/numerol- ogy2.html), a site that sells "powerful sacred jewelry"
5. "Meanings of Numbers" (mysticalnumbers.com), a site dedicated to the symbolism of numbers

After that we get more angels, more numerology, more biblical revelations, etc., up until hit twenty or so, in which we get a grammar lesson on the word "number."

I shall stick with my less exciting definition throughout, ignoring this flood of misinformation that thrives online as a consequence of a condition that affects very large segments of the population: innumeracy—lacking basic knowledge of arithmetic and elementary mathematics.

The origins of counting are lost in deep prehistory, although one can imagine its utility for primitive hunter-gatherer groups: How many people should go out on the hunt? Did everyone make it across the swollen river? The first abstract number systems that used symbols, including placeholders (e.g., systems in which a digit's location also conveys meaning), was the Mesopotamian sexigesimal (base 60) system for which we have records from 3400 B.C., and the Egyptian base 10 system from a few centuries later. The earliest applications of number are generally posited to be for reckoning the seasons and for business purposes.

Eventually, the abstract symbols became the subject of study themselves, and a philosophy of numbers developed that we now call mathematics. The so-called "natural" numbers 1, 2, 3, 4, . . . , the ones used for counting, were augmented by incorporating first zero, then negative integers, rational numbers (fractions, the ratios of integers), and real numbers, some of which have decimal fractions that never repeat and so cannot be represented by the ratio of integers; these are dubbed irrational numbers and include pi and e, the base of the natural log system. The emergence of geometry, algebra, and calculus both transformed mathematics into a major area of intellectual inquiry and led to the discovery of numerous applications in the study of the physical world we call science. Galileo stated the case for mathematics powerfully:

The universe cannot be read until we have learned the language and become familiar with the characters in which it is written. It is written in mathematical language, and the letters are triangles, circles and other geometrical figures, without which means it is humanly impossible to comprehend a single word. Without these, one is wandering about in a dark labyrinth.[2]

In defining the limits of our perceptions and the enormous extensions thereto that science provides, I used a lot of numbers. The following chapters are suffused with numbers as well. Almost without exception, however, these numbers are used in their simplest sense of counting and measuring. Whether on a graph in which we express the relative distances of numbers from each other, or in probability, where we simply count the number of outcomes of interest divided by the total number of possible outcomes, or in correlation analysis in which we are simply quantifying how far from a representative model line various measurements fall, we are using the numbers to count and to measure quantities. No math phobias need be stirred up—it's all just about counting and measuring real-world quantities.

IGNORE NUMBERS AT YOUR PERIL

Without numbers, the best of intentions can go awry. One of my favorite recent examples comes from an analysis of energy generation in Colorado by the engineer Henry Petroski. By 2013, Colorado proudly proclaimed that they had achieved their goal of generating more than 10 percent (13.8 percent to be precise) of the state's electricity needs from zero-emission wind power. Sounds laudable, yes? The problem, Petroski pointed out, was that the net effect of this expensive achievement—a lot of wind power on the Colorado grid—actually increased the state's carbon footprint, the amount of CO_2 it emits into the atmosphere annually.

Wind is, of course, a variable source of power. To keep the electrical grid supply in balance with its demand, other sources of electricity must be added in an amount inversely proportional to the wind-generated power. The principal alternative source in Colorado is coal-fired power plants that produce far more CO_2 per kilowatt-hour generated when they are turned on and off frequently than when they run at a steady rate. Thus, in fulfilling a goal designed to mitigate its impact on the global climate, the absence of a quantitative analysis of its energy production system meant Colorado achieved the opposite effect. Numbers matter.

NUMBERS WITHOUT CONTEXT AS MISINFORMATION

Just as ignoring numbers can lead to flawed decision-making, reifying numbers without putting them in context can be equally problematic.

I was once at a Columbia College alumni event at which the dean was asked if it were true that Columbia limited the number of students who could graduate with Latin honors to 20 percent of the class. "Yes," the dean replied. "We think it is important to highlight the achievement of our truly outstanding students." The alum protested vigorously: "That's doing a real disservice to our graduates. When I am reviewing two applicants, one from Columbia and one from Harvard, and the one from Harvard graduated magna cum laude and the one from Columbia has no such distinction, I'll take the Harvard applicant every time." Left unmentioned in the alum's analysis was the fact that 93 percent of Harvard graduates that year received Latin honors. Thus, virtually the only information contained in Latin honors at Harvard is that the student was admitted to Harvard. Ascribing significance to this datum as an indicator of performance or future success is meaningless.

Even professional admissions officers evidently engage in such fallacious reasoning. In an article published in *PLOS ONE* and quoted in the *Wall Street Journal*,[3] the authors reported that when twenty-three admissions officers from business schools across the country were asked to evaluate nine fictitious applications from undergraduate schools of comparable quality but with very different grading standards, the applicants with the higher GPAs were admitted more frequently, even when the admissions officers acknowledged that students from schools with tougher grading standards had to work harder to achieve their (lower) GPAs. In a related study, the applications of 30,000 real students seeking entry to elite business schools were analyzed, and those from the schools with the highest mean GPA for all students (i.e., those schools at which grade inflation was most rampant) had the highest probability of admission.

To a mathematician, the number 2 and the number 3 need no context; summing them will always yield 5, whether they represent oranges or supermassive black holes. To a scientist—or to any person using numbers to make decisions in the real world—context matters. Five oranges weigh less and taste better than five supermassive black holes. A GPA is a number (now, absurdly, often calculated to four or five significant figures). But it carries no meaning without context—at minimum one needs the mean GPA of a student's fellow graduates and, ideally, an examination of the rigor of the courses taken and the grading standards for each to impute any meaning to a GPA.

When numbers are large, the lack of context often renders them completely bereft of meaning unless one does a quick calculation to measure their relevance. Three examples make this point:

1. In January 2013, President Barack Obama announced, in response to the Newtown school massacre, a raft of mental health measures to combat gun violence. The expenditure sounded enormous: $155 million. That amounts to 50 cents per person or 52 cents per gun—or about 8 seconds of psychiatric care per person or enough gas to drive 3 miles toward a clinic appointment.

2. In 2012, Walmart made a big splash by announcing that it would buy $50 billion of US-made goods over the next decade. Since it spent $260 billion in 2012, assuming a low inflation rate of 2.5 percent per year, this amounts to a "made in the USA" label on 1.5 percent of all its purchases by 2022.

3. In July 2014, the Obama administration's State, Local, and Tribal Leaders Task Force on Climate Preparedness and Resilience issued its recommendations, and Obama traveled to Missoula, Montana, to announce major initiatives such as $13.1 million to develop maps to help cities plan for weather-related disasters and $10 million "to train tribes on climate change adaptations."[4] To place these numbers in context, $10 million represents what the federal government spends in about 81 seconds. And the cost of flying the president out to Missoula and back on Air Force One to make the announcement? About $2.5 million.

Context matters, too.

"RELIABLE" SOURCES

Alexander Pope's dictum "to err is human" requires that one never put away the innumeracy detector. In a 2014 article from Sigma Xi's daily digest of science news, we find the following:

> Ocean thermal energy conversion, or OTEC, is the focus of several projects underway around the world in an effort to supply Earth with an abundant source of renewable energy—essentially creating an ocean-fueled steam engine. Projects are looking at more efficient delivery methods, using ships that would search the ocean for areas with the best temperature ratios. The ideal temperature difference between deep ocean water and that on the surface needs to be at least 20 degrees Celsius, or 68 degrees Fahrenheit.[5]

Now it is true that a temperature *measurement* of 20°C is equivalent to a temperature measurement of 68°F. But it is definitely not true that a temperature *difference* of 20°C equals a temperature *difference* of 68°F (it is actually 36°F). Indeed, the putative ocean thermal energy conversion process could never work anywhere if it required a temperature difference of 68°F; because the deep ocean temperature is a few degrees above freezing (32–40°F), that would require a surface temperature in excess of 100°F, which occurs nowhere on Earth (with the possible exception of brief periods in very shallow tropical seas where there is no deep ocean water to use as a heat exchanger).

MILLION, BILLION, SHMILLION . . . WHATEVER

When the numbers are big, the room for error is ever greater, and errors are rampant. The following paragraph appeared in the January 21, 2013, edition of *Canadian Business* in an article about Harry Winston's diamond business:

> The Toronto-based miner recently announced it was selling its luxury retail business to Swatch Group for US$750-billion, plus the assumption of up to US$250 million in debt. The decision to sell the retail arm of the business to focus on mining follows Harry Winston's move late last year to acquire BHP Billiton's 80% stake in Ekati for $500 million, Canada's first and largest diamond mine.[6]

One might think that a magazine devoted to business readers would actually care about the difference between a million and a billion. The fact that neither the author nor the editor noticed a problem here is exemplary of the endemic lack of quantitative reasoning that surrounds us. First of all, context aside, the numbers presented don't even make sense. If Harry Winston were actually selling its retail business to Swatch for $750 billion, why would the $250 million in debt even be worth a mention—it represents a 0.03 percent difference in the price. And could anyone possibly think that Harry Winston's luxury retail business was worth 20 percent more than the GDP of Switzerland? If it were, and Swatch could afford it, why didn't they just buy the whole country instead and save $120 billion—the "change" would be enough to purchase Boeing and Goldman Sachs outright as well, so Swissair could have free planes, and tax shelters for rich Americans would be easier to come by.

The use of million, billion, and even trillion as false synonyms can only be described as rampant (the imported food example in chapter 4 is another example). For people of my age and ancestry (my mother was British), this is almost excusable, because decades ago a "billion" in the United Kingdom was a million million (or what we now call a trillion = 10^{12}; a trillion was a million million million = 10^{18}). Italy and France still use this scheme, the latter, curiously, switching from the current American scheme back to the old British scheme in 1961; they refer to 1,000 million as a milliard. But the United Kingdom officially changed to the American usage in 1974, and this is now the standard in all English-speaking countries. So no excuses for anyone under 50, and everyone else should get with the program.

NUMBERS MATTER

Don't succumb to number numbness. Granted, big numbers are not always easy to picture (although it is *always* possible to construct a useful analogy). But each number has a specific meaning, and that meaning helps us to understand the world. Think of it in terms of money if necessary: a million pennies in your bank account is $10,000, but a billion pennies is $10 million. Presumably, you'd notice that difference. Numbers deserve respect; they are the currency of a rational, scientific view of the world and an important shield in the Misinformation Age.

4
DISCOVERIES ON THE BACK OF AN ENVELOPE

I t was the bottom of the seventh inning about twenty-five years ago at Shea Stadium, and Dwight Gooden was pitching a masterful game for the New York Mets. Because of the Mets' woeful offense, however, it was still a scoreless tie. Then it started to rain.

I was at the game with my brother-in-law Jeffrey, whose interest in baseball I never quite understood. As a lifelong Red Sox fan, I had a clear understanding of my own engagement. Red Sox fandom always served for me the same purpose as Greek tragedy—one always knew the outcome, yet one experienced catharsis getting there nonetheless. The past decade, with three World Series championships (after eighty-six years with none) has actually detracted somewhat from that experience; when the heroes regularly triumph, its gets to be more like a soap opera than Greek tragedy—far less interesting.

But Jeffrey is more of a theoretical fan than a passionate one, and his first response to the rain was not that this wonderful pitching performance might go for naught, but instead: "How much do you suppose the ball slows down as it encounters the raindrops? Does it significantly affect his fastball for the hitters?"

These may well be questions no one has ever thought to pose before. One is unlikely to find the answers in a textbook or on the Internet. Indeed, an Internet search with key words "fastball," "slows," and "raindrops" produced only the following:

"TANYA: Just go slow, think about what you're doing.
The crafty Limbor smoked a couple of fastballs. Oakly (glances at the
    darkened sky as raindrops . . . "

Not much help from Google for the baseball question, however.

My mustard-stained napkin was considerably more useful, because it is possible to calculate an answer—or at least give a reasonable estimate of the size of the effect. Right; geeks are like that, I hear you muttering. But the point is that *you* can do it too. In fact, estimating in this way—a back-of-the-envelope or napkin calculation—is one of the easiest of the scientific habits of mind to acquire, because it requires no more than arithmetic and a dash of confidence. I will assume here you can do the former. I hope to foster the latter with a series of examples.

QUALITATIVE QUANTITATIVE REASONING

Science is often portrayed as a rigidly precise discipline in which the sixth decimal place may reveal a critical cosmic truth. Indeed, sometimes that's the case: a minuscule discrepancy in the position of Mercury in its orbit around the Sun mentioned earlier led to the first confirmation of Einstein's revolutionary theory of general relativity,[1] whereas measurements of tiny fluctuations in the cosmic background radiation provide critical clues to the universe's first moments. But focusing on precision is not always necessary, and doing so often fails to capture the spirit of the scientific process. Scientists do quantify Nature, but often in a more qualitative way.

Indeed, one of the distinguishing habits of a scientific mind is the ability—and the willingness—to make rough estimates of unknown (and often unknowable) quantities. We call these "back-of-the-envelope" calculations because they almost always contain few enough steps to fit on the back of an envelope (or, more frequently these days, a napkin). The idea is not to calculate something precisely, not to spend hours searching the library or the Web for information, but to get a quick, rough idea of how big, distant, heavy, damaging, or expensive something is.

Scientists often use this process to explore the feasibility of undertaking some set of observations, to design an experiment, or to evaluate the tractability of a computer simulation. However, it is also often useful in assessing the plausibility of statements one hears or reads, in contextualizing sensationalized news stories and political claims, or simply evaluating one's own life experiences.

I also understand from former students that back-of-the-envelope reasoning is a skill highly valued in lucrative "management consulting" jobs (just in case you need a crass, practical reason to be interested in finishing this chapter). Indeed, a few years ago, right after one of my students finished interviewing with a leading management consulting firm, she popped into my office in a very excited state. "Professor Helfand!" she said, almost out of breath. "I just had my interview, and you'll never guess what they asked!" I didn't have a clue. "The first question was 'how many fax machines do you think there are in Brooklyn?' It was just like your back-of-the-envelope questions, only easier. They asked several more and I just blew them away." This scene has been repeated numerous times in subsequent years, so it seems fair that I advertise this chapter as offering a useful skill.

DESENSATIONALIZING THE NEWS

I frequently use envelopes and napkins to debunk, or place in perspective, sensationalized news stories. For example, every decade or so the media gets excited about "killer sharks." By the beginning of the fall term a few years ago, despite weeks of headline coverage concerning the "shark menace," precisely two people had died in the United States from shark bites since the preceding January. What fraction is that, you might ask yourself, of all the people who have died during the year in the United States? The answer can be easily determined as follows.

As of February 22, 2015, at 1:30 PM EST, the U.S. Census Bureau's "population clock"[2] claimed there were 320,391,554 people in the United States. This is just the kind of number we do *not* need for any back-of-the-envelope calculation. First of all, it's silly—there is no conceivable mechanism by which we could know, to the nearest person, how many individuals are alive in the United States at any given moment. Second, we are trying to make an estimate accurate to at best a factor of two, so at least seven of the nine decimal places in this number are irrelevant. Rather, the calculation should proceed as follows.

There are about 320 million (3.2×10^8) people in the United States; the average life expectancy is about 78 years (averaging men and women—note that average life expectancy is just what we want here because it tells us how long the average person lives). This means that 3.2×10^8 people/78 years = 4.1×10^6 people die each year.

By early September, the year is about 245/365 days (67%) over, so roughly $0.67 \times 4.1 \times 10^6 = 2.7 \times 10^6$ people will have died by the time the semester begins. The shark victims, then, represent less than one out of every million deaths. Not exactly a major health threat. In contrast, two U.S. residents die every 130 seconds from smoking cigarettes,[3] and two die every 32 minutes in car accidents.[4]

Notice that in this kind of calculation one does not worry about the fact that there are not exactly the same number of people in each of the 78 years of life, that the death rate might vary slightly from month to month over the year, or that the U.S. population is not exactly 320 million. My interest was in the relative fraction of deaths from sharks—was it 1 in 100 or 1 in a million? Clearly, it is closer to the latter, and whether it is 1 in 1.21 million or 1 in 0.98 million doesn't matter at all in getting a feeling for the relative importance of the problem.

Note also that the input numbers for such calculations are rarely accurate to more than one or two significant figures (320 million people, 78-year life expectancy). In keeping with the rules I shall enunciate in chapter 7, one should never quote the answer from such a calculation to more than two significant figures, although it is fine to carry along more decimal places on your calculator until you finish calculating.

Another favorite recurring news story concerns forest fires in the West. In some years it is Montana and Idaho; in others it is California or Colorado. In each case, however, one is left with the distinct impression that a large fraction of a state is burning: "Fires Rage Across Colorado," headlines reported in 2002. Tourism dropped to zero—who wants to camp in the middle of a burned-out forest?—and the governor had to call a press conference to say that the entire state was not burning (the press conference, of course, got much less coverage than the fires).[5]

The sensationalism lavished on these types of stories has two characteristics. First, it ignores both history and context (there are—and for healthy forests must always be—many fires every year); second, it quotes numbers in inflammatory (pun intended) ways—they must be *BIG* numbers. Fires are always reported as the number of acres burned. Do you know how big an acre is?

If not, you are to be forgiven, since there seems to be some confusion on the subject in the English-speaking world (the only portion of the world benighted enough to still abjure metric units). The word comes from the Latin *agrus*, which means a plowed or open (but distinctly not forested) field. In the United States and United Kingdom, an acre is defined as 160 sq. rods—which probably doesn't help much unless you happen to know that a rod (also known as a perch) is 5.5 yds. That makes a U.S. acre 4,840 sq. yds. or 43,560 sq. ft. In Scotland, however, an acre is 6,150 sq. yds., and in Ireland it is a whopping 7,840 sq. yds. Is this absurd or what? In passing, it is worth noting that the metric unit for land area is the hectare, a neat 100×100 m = 10^4 m^2 (or roughly 2.5 U.S. acres).

In any event, during 2002 in Colorado, we were breathlessly told that 500,000 acres had burned. That sounds like a huge number. It makes more sense, however, to quote the number in square miles (or even better in square kilometers): 500,000 acres is less than 790 sq. miles. The total area of Colorado is 2.7×10^5 sq. miles, so roughly 0.29%—less than one-third of 1 percent—of the state actually burned in one of the worst fire seasons on record for Colorado. If you have a 2,100 sq. ft. house, this is equivalent to a fire in a 2 ft. × 3 ft. trash bin. Assuming you put out the trash fire (as they did the fires in Colorado), you'd be unlikely to report to friends and family that your house had burned down (or that 557,000 sq. mm of your house had burned).

QUESTIONING AUTHORITY

Envelopes can also be used to spot errors in apparently authoritative sources. (As you can see, I intend to continuously assault your habitual, passive acceptance of [mis]information). I once had a student who subsequently became an editor for *Foreign Affairs*, the publication of the Council on Foreign Relations and one of the most authoritative journals in its field. By using the back-of-the-envelope calculational techniques he had learned in my class, he frequently caught errors in articles written by UN officials, former secretaries of state, and others. This skill is not just useful for scientists and management consultants.

Some years ago in the *New York Times* (my favorite place for catching errors given its self-proclaimed status as the "paper of record"), I read an article on the dangers of inadequately inspected food flowing into the United States from foreign countries. It stated that "Although that vast majority of the 30 billion tons of food imported annually is wholesome . . ."[6] This number prompted me to do a little calculation.

Consider that 30 billion tons = 30×10^9 tons $\times$ 2,000 lbs./ton = 6×10^{13} lbs. of food. At the time, there were roughly 300 million people in the United States, and most of them ate (in fact, most apparently ate too much). But if we all ate only imported food, we would need to consume 6×10^{13} lbs./year/3×10^8 people/365 days/year = 550 lbs./person/day.

Although we certainly waste a lot of food in this country, and are, as a nation, obese, no one eats 550 lbs. of food a day. Clearly, this number must be wrong by a factor of at least 1,000. So much for the paper of record.

If one is alert, one finds such errors everywhere. Here are a couple more examples collected over a four-week period a couple of years ago:

From *Talk of the Nation* on National Public Radio:

> HOST: So today we are joined by elevator historian and associate dean of the College of Arts and Architecture at UNC Charlotte, Lee Gray. He joins us by phone from his home in Charlotte. And welcome, Lee Gray, to *Talk of the Nation.*

A few minutes into the conversation, the host and Dr. Gray were discussing the fast elevators installed in modern skyscrapers:

> DR. GRAY: Well, a lot of that depends on the experience you're looking for. Certainly, observation elevators, you know, whether it's the Eiffel Tower or similar buildings where you can see out as you're riding up, you know, offer extraordinary experiences. And then you have—on the one hand, they seem normal elevator rides but something like, you know, some of the world's fastest elevators

that—in Taipei 101 in Taiwan, for example, goes about 3,300 feet per second. And it's moving so fast that the elevator car is some-what—is lightly pressurized just like an airline cabin. Otherwise, you would suffer discomfort in your ears. And some of the fast elevators also have what we might think of as speedometers in them.[7]

If Taipei 101 did have a speedometer, I bet it wouldn't read 3,300 feet per second—that's 2,250 miles per hour or about 4 times the speed of sound; you'd get from the ground floor to the roof of this 101-story building in less than half a second. Assuming the elevator reached this speed over the first 10 floors, the acceleration you'd feel would be 1,150 times the Earth's gravitational acceleration, giving you an apparent weight of about 170,000 lbs. and crushing you into a puddle on the elevator floor.

It's clear the guest meant to say 3,300 feet per minute (which is still pretty fast!), and, of course, anyone can make a slip like this. But the fact that neither the host, the person who placed this on the website, nor the editor noticed anything untoward in the idea of an elevator traveling at four times the speed of sound is, to me, a classic example from a nation of innumerates accepting whatever they hear.

Without one's numerical radar always turned on, it is easy to be misled by someone—a journalist, a politician, or even an obstreperous neighbor—trying to make a point. Take, for example, what Ryan Scott, chief executive of Causecast, a nonprofit that helps companies create volunteers and donation programs, was quoted as saying in a *New York Times* article from June 15, 2014.[8] Comparing the level of philanthropy at Apple in the wake of new chief executive Tim Cook's charitable initiatives relative to what other companies such as Microsoft were doing at the time, Scott said that, although off to a good beginning under Cook, Apple's philanthropic goals "could be much higher" based on its talent and resources. The article then went on to say:

By comparison, Microsoft says that it donates $2 million a day in software to non-profits, and its employees have donated over $1 billion, inclusive of the corporate match, since 1983. In the last two years, Apple employees have donated $50 million, including the match.

Those stingy Apple employees! But wait. Let's do the comparison carefully. Both employee donation amounts include a match by their corporate parents, so that's fair. However, the Microsoft total is integrated over thirty-one years, whereas the Apple amount is summed over two years. On an annual basis, that's a total of $32 million from the Microsoft employees and $25 million from the employees at Apple. And in 2014, Microsoft had 99,000 employees and Apple had 80,300. The final result then is that Microsoft employees, on average, contributed (with the match) $323 each, whereas Apple employees contributed $311 each, less than a 4 percent difference. Not so stingy after all.

THE FAMOUS FERMI PROBLEM

One of the principal expositors of back-of-the-envelope calculations was Enrico Fermi, the great Italian physicist who, after having fled the Fascists, ended up at Columbia University in 1938 and left three years later when President Nicholas Murray Butler would not let him drain the old swimming pool to install the world's first nuclear reactor. (Fermi moved to Chicago and built his reactor under their football stadium instead—have you heard much about the University of Chicago's football team lately?) Indeed, in honor of his penchant for solving such estimation problems in his head even when only the most minimal information was available, these problems are often called Fermi problems. Without doubt the most famous of these is, "How many piano tuners are there in New York?"

Although this clearly is not a matter of burning urgency to our nation's geopolitical success or to your own personal financial future (unless you aspire to becoming a tuning wrench salesperson), it is

1. a factoid you, or anyone you know, is extremely unlikely to know off-hand and
2. a good example of how different people will arrive at similar numbers while making a calculation that requires several steps and a number of assumptions.

I have posed this problem to many of my classes and have never failed to be impressed that novice calculators all come up with very similar answers. To tackle this (or any such) problem, begin with what you know (or can easily guess or find):

Population of New York: 8 million
Time to tune a piano: ~2 hours
Frequency of tuning: ~1 per year
Working days in a year: 365 − 104 (weekend days) − 15 (holidays) − 20 (vacation) = 226 days

Note that the accuracy of these numbers varies. The population is accurate to better than 10 percent, and the number of days worked per year for the average person is even more accurate. The time to tune a piano probably ranges from less than an hour (although we might add a little travel time) to many hours (before a major concert at Carnegie Hall), and the frequency of tuning is even more variable and uncertain (probably once a week at major concert venues and less than once per decade for the piano of an elderly couple in Brooklyn). Nonetheless, these numbers serve as rough averages for the average piano—I would guess.

Next, estimate what you don't know: the number of pianos per person in New York. Since the U.S. Census Bureau does not ask about pianos, there are no official data. There is little doubt, as with almost any question you can dream up, that you can find an answer on the Web, but how will you know if the number you find there is valid? I performed a Google search for "number of cars in the U.S." and, on the first page of hits, found answers that differed by twenty-six million—despite the fact that the reported numbers were quoted to between three and nine (!) significant figures. Without being able to do a rough calculation yourself—without the self-reliance this brings—you are a dependent creature, doomed to accept what the world of charlatans and hucksters, politicians and professors provides, with no way out of the miasma of misinformation.

How do I estimate the number of pianos in New York? Let's start with the extremes. Is it zero? Categorically no, since I have a piano and I live in Manhattan. Is it eight million? No again, since my wife and I share one. Furthermore, I know most of the people in our building (approximately 45 apartments with an average of 2 or 3 people each, so maybe 100 people), and I only know of one other piano. Maybe there are two or three that I don't know about. But my building, occupied mostly by Columbia affiliates, is clearly not typical of New York City as a whole; my building's mean income is almost certainly well above the average for New York, and it is reasonable to assume that piano ownership and income are correlated. I'd guess the true value is around 1 percent of the population, or 80,000 pianos—maybe we'll round up to 10^5 just to be generous.

Wait! you might say. What about all the concert halls? And schools? And universities? This provides an important lesson in relevance, again using the envelope. How many concert venues are there in New York? 100? 200? And suppose they each have several pianos—so that's an extra 500, or maybe 1,000 pianos at most—less than a 1 percent correction to my estimate. And schools? Each school has one, or a few, pianos, yet each school has hundreds to thousands of kids, and school kids make up only

---

> **BOX 4.1 ESTIMATING A SPECIFIC POPULATION**
>
> Why do I say school children make up 12/78 of the population?
>
> Imagine everyone lives to the age of 78. Clearly each person spends just as much time between ages 0 and 1 as between ages 77 and 78 (i.e., all years are the same length, and you can't skip any).
>
> Most people spend 12 years in school grades 1 to 12. Thus, at any given time, roughly 12/78th of the population, or 15 percent, is in school.
>
> In reality, the birth (and immigration/emigration) rate is not precisely constant in time, and of course everyone doesn't live precisely 78 years and then drop dead. The U.S. Census Bureau's actual tally of the population by age[9] gives (in July 2014) 49.66 million between ages 6.0 and 18.0 = 15.6 percent of the population. So 15 percent isn't such a bad estimate.

---

about 12/78 of the population (see box 4.1), so again, a small correction. Columbia University has 38 pianos (I asked) and 23,000 students. Again, this is less than a 1 percent effect.

Because there is limited space for calculating on the back of an envelope, it is essential to avoid going off on irrelevant tangents and pursuing trivial corrections to your rough estimates. Remember the goal: to obtain an approximate number, good to a factor of two or even a factor of ten; in service of this goal, all minor effects can be safely ignored. So, 10^5 pianos × 1 tuning/year × 2 hours/tuning × 1 day/8 hours × 1 year/226 days = 111 tuners. Now, of course, it is not exactly 111—as I said, quote back-of-the-envelope answers to one or two significant figures. In this case, more than one would overestimate the accuracy; thus, I'd say "about 100." Many of them may be part-time tuners, so maybe a couple of hundred. But it is highly unlikely that there are thousands (unless they are all very hungry), and it is highly unlikely there are only ten (they would each have to tune thousands of pianos a year).

In many years of asking this question, my students always (1) think I'm weird and say "How can I possibly know that?" and then (2) go home and come up with a number between 50 and 500. That is, they find that the order of magnitude (see box 4.2) of the answer—about 100—and that is all we wanted to know. And then at interview time they respond to a management consultant's question about the number of fax machines in Brooklyn the same way and garner an attractive starting salary.

BOX 4.2 ORDERS OF MAGNITUDE

The "order of magnitude" of a number is a rough estimate of how big it is. One order of magnitude is a factor of ten.

Two orders of magnitude is $10 \times 10 =$ a factor of 100. Five orders of magnitude is $10 \times 10 \times 10 \times 10 \times 10 =$ a factor of 100,000.

If your checking account is worth $12, and your boss decides to confer on you a bonus that increases the money in your account by about two orders of magnitude, you will now have about $12 \times 10 \times 10 = \$1,200$— or it could be $1,100 or $1,350 (we'd still say your bonus increased your net worth by a couple of orders of magnitude).

If you have a box containing 200 raisins and you eat 178 of them, you have decreased your stash by a factor of $200/(200 - 178) = 200/22 = 9.09$—or about one order of magnitude.

The number 98 and the number 212 are both "of order" 100; that is, they are not literally equal to 100 (obviously), but for some purposes they are close enough to 100 to be considered the same.

Order-of-magnitude calculations are useful in estimation either because details are unnecessary or because they are literally impossible to obtain. The number of piano tuners in Manhattan is an example of the former case; the number of stars in the universe is an example of the latter.

ESTIMATING THE UNKNOWABLE

In the third-century B.C.E.,[10] Archimedes wrote about the number of grains of sand needed to fill the universe. Clearly, he had not counted the grains of sand on Earth or measured the size of the universe—it was just a poetic way of saying the universe is a big place. Archimedes' speculation has subsequently been paraphrased as saying the number of stars in the sky is greater than the number of grains of sand on Earth. How does this wild estimate comport with modern knowledge?

Telescopes now extend our vision to stars four billion times fainter than the stars Archimedes could see with his naked eye, but we are still very far from being able to count them all. So we estimate that the number of stars within 50 light-years of the Sun (these we can count) = 1,000 and that the fractional volume of the galaxy this 50-light-year sphere represents = 10^{-8}, which implies about 100 billion (10^{11}) stars in the Milky Way.

The number of galaxies in the visible universe comes from taking the number we count in the deepest picture of space ever taken, the Hubble Space Telescope's Ultra-Deep Field, which covers a piece of sky a little less than 2 percent the size of the full Moon.[11] This image includes a few dozen stars in our Milky Way and roughly 10,000 distant galaxies. We then multiply this many galaxies by the number of such patches it takes to cover the sky (about thirteen million): again, curiously, but purely coincidentally, the number is about 10^{11}, a hundred billion galaxies. So the number of stars in the observable universe (see box 4.3) is roughly 10^{11} galaxies × 10^{11} stars per galaxy = 10^{22}. This is, strictly speaking, an "unknowable" number in that no one can, by dint of stubborn persistence, go out and count them. First, 99.99999999999 . . . percent of them are too faint for even our largest telescopes to record individually; second, if we set all 7.2 billion people on Earth to work on this task 12 hours a day, 7 days a week, and each one counted 1, 2, 3, 4, 5 . . . getting to 20 every 5 seconds, it would take 22,000 years to finish the count (far longer than we have had numbers to count with).

BOX 4.3 WHAT IS THE OBSERVABLE UNIVERSE?

Light does not travel instantaneously between points in space. It has a finite speed of 3×10^8 m/s (or about 1.1×10^9 km/h—flying at this rate you could get from New York City to Tokyo in about 1/30th of a second).

Since light takes time to travel, we never actually see the current moment. Looking down at your hand, you do not see it as it is right now, but rather as it was a miniscule fraction of a second earlier. Now, this interval is so small, given the short distance between your retina and your hand, that the difference is utterly negligible. The discrepancy becomes significant, however, when exploring much greater distances. One light-year is the distance light travels in one year (about ten trillion kilometers). If you look at a star 500 light-years away, you are seeing it as it was 500 years ago. Thus, the deeper you peer into space, the further you are seeing back in time. If this star had exploded 499 years ago in a spectacular event called a supernova, we would not—indeed we could not—know about it. We wouldn't know that it happened until one year from now.

Indeed, any event that happened beyond a certain point in the past is unknowable to us if the signal from it hasn't had time to reach us. It is not that our telescopes are too weak or that our instrumentation is insensitive. We simply do not yet have access to the information—no matter how rapidly you read, you'd be hard-pressed to read a friend's e-mail if it had yet to arrive in your inbox.

As a consequence of this limitation, astronomers refer to the observable universe as the volume of space from which we are, in principle, able to record light. This is the distance from which light has had time to reach us since the universe began 13.78 billion years ago (by definition, then, a distance of 13.78 billion light-years in every direction). We are quite confident that the universe is, in fact, much larger than this, and every day we see another light-day out into space, every year another light-year.

However, this is a number astronomers need to know if, for example, we are to develop an understanding of the distribution of the elements in the periodic table. Because stars cook up all these elements in their cores and disgorge them at the time of their deaths, the number of stars that make elements is essential in understanding why platinum is rarer than gold and why both are rare compared to iron. We can compute this useful number on the back of an envelope from the single long-exposure image taken of the Hubble Ultra-Deep Field.

The number of grains of sand on all the beaches of the world is a less useful number. If you'd like some practice, however (and gain another new perspective on the scope of the universe), try figuring it out.

SANITY CHECKS

I once attended a conference at which scientists from many disciplines gathered for a weekend of lectures and conversation. After a talk by the discoverer of many extrasolar planets—planetary bodies that orbit parent stars other than the Sun—an eminent biologist from a prestigious university asked how many of these systems he thought we'd be able to visit and explore with spacecraft. We astronomers are used to dealing with "astronomical" numbers when it comes to distances, but apparently other scientists are not. The speaker let the questioner off gently, noting that after traveling thirty-six years, the *Voyager 1* spacecraft became the first manmade object to leave the solar system in 2013; it crossed the boundary 18.7 billion kilometers from Earth. The nearest star system, Alpha Centauri, may or may not have some planets to visit (a matter of some controversy at the time of this writing), but it is 4.37 light-years away, and 1 light-year is 9.5 *trillion* or 9,500 billion kilometers. The very nearest option, then, is more than 2,200 times farther away than the distant *Voyager 1* has traveled in thirty-six years.

Had the questioner scribbled the following on his program before speaking, he would not have asked the question:

> Nearest of the stars with planets discussed in the talk: 10 light-years
> Velocity a spacecraft must achieve to leave Earth: 11 km/s (and we
>   can't make them go much faster than twice this value yet)
> Speed of light: 300,000 km/s
> Spacecraft speed = 11 km/s/300,000 km/s = 3.7×10^{-5} light-speed
> Time to reach closest planet: 10 light-years/3.7×10^{-5} = 2.7×10^5 years
>   (or about twice the time that *Homo sapiens* has been around)

It's not that hard to avoid embarrassing questions.

PROVIDING CONTEXT

It is common to be confronted with a number for which one has absolutely no context; without context, the number is virtually bereft of meaning. In this situation there are two approaches one can adopt: (1) ignore the number or (2) clothe it with context. Most people make the former choice; a scientific mind strives for the latter. The U.S. national debt is over $17.5 trillion; no one I know has a visceral feeling for what a "trillion" means.[12]

The national debt becomes an abstract concept devoid of real meaning—so we ignore it. A simple calculation, however, gives it meaning: 17.5×10^{12}/3.2×10^8 citizens = $55,000 per citizen or $220,000 for a family of four. Is everyone as comfortable with that?

Part of this debt was accrued building nuclear warheads, of which we now have—after dismantling a large number of them, also at great cost—approximately $4,800. These warheads have a very broad range of explosive yields, but a median of one megaton (roughly fifty times the destructive force of the atomic bombs dropped on Hiroshima and Nagasaki[13]) is a plausible guess. That works out to about 1,300 lb. of dynamite for every person on Earth; 1 lb. is enough to blow up a car and kill everyone in it. Do we really need a new manufacturing facility for nuclear warheads?

Another part of the debt was accumulated buying foreign oil. We currently use about 20 percent of the total energy the world consumes. We have $3.2 \times 10^8/7.2 \times 10^9$ (4.4 percent) of the world's population. How much more energy would be needed if the whole world was using energy as we do? The answer is 450 percent.

PROVIDING PERSPECTIVE

Having taught at Columbia University in Manhattan for nearly four decades, I have fielded innumerable (well, not literally, but I have not bothered to count) questions from prospective students' parents worried about the safety of their little darlings in big, bad New York City. I've produced these calculations just for them (and for any of you out there who have been worried about a visit to the Big Apple).

In 2014, the last year for which complete statistics are available, New York City had by far the lowest murder rate of the ten largest cities in the U.S.: 3.8 per 100,000 residents, or less than one murder per day. Using the calculation we employed earlier, roughly 8.4×10^6 people/ 78 years (life expectancy) = 108,000 people die each year in New York City, the chances of dying from a violent act are roughly 0.3 percent. Given that 80 percent of murder victims know their murderers, the chances of dying in a random act of violence are considerably less than one-tenth of 1 percent. Of course, I tell these parents that they could send their child to school in Riverside, California (which had a murder rate the same as New York's in 2012), and let them have their own car on campus (not allowed in New York). Deaths among 18- to 22-year-olds from car accidents are 19.2 per 100,000 in 2010,[14] or 25 times higher than being murdered by a stranger.

These parents are also concerned about the dangers of the subway system. For this I send them to chapter 6.

CONCLUDING THOUGHTS

If your credit card bill arrives with a total due that is $10 more than you expected, you probably wouldn't think twice about paying it. If the total was $10,000 more than you were expecting, you might be more than a little upset. That's a factor of a thousand more! Well, a billion is a factor of a thousand more than a million, but few people seem to care or notice.

Scientists, however, think numbers matter. If the Greenland icecap were to melt, would the oceans rise seven millimeters or seven meters? It matters (and the answer is a little over seven meters or about twenty-three feet). Does the smallpox vaccine induce a severe reaction in one in a million of the soldiers ordered to take it or one in a thousand? It matters. Are we destroying the rainforests of the world at 1,000 hectares per year or a million hectares per year? It matters.

The endemic misuse of numbers is a characteristic of the Misinformation Age. Technology has now put more misinformation at your fingertips than has existed in all of human history. One tool you have to combat the misinformation glut, to make sense out of nonsense, is the back of an envelope.

Now, as for those raindrops slowing the baseball, see box 4.4. And remember: if you are ever at a baseball game with me, hold on to your napkin.

BOX 4.4 BASEBALLS AND RAINDROPS

As is often the case with back-of-the-envelope problems, there are a number of ways to approach this. Here's the one I used.

The pitcher's mound is 60 feet, 6 inches from home plate (18.44 m). I guessed a baseball was about 3 inches in diameter and weighed about 6 ounces. I subsequently looked it up on the Web and found that the National Collegiate Athletic Association (NCAA) weight requirement for a baseball was 5.12 ± 0.035 oz., with a circumference of 9.05 ± 0.05 in. (I was delighted to see that the NCAA recognizes the concept of measurement and manufacturing uncertainties!) So my guesses on my napkin were within ~15 percent in both cases, but I will use the real numbers since I now have them (5.12 oz. = 145.2 g, and 9.05/2π implies r = 3.66 cm).

Now, it was raining fairly lightly, and I estimated it would take at least four or five hours to accumulate an inch of rain at that rate. I chose 0.5 cm/h to use in the calculation. This means that in an hour, a layer of water 0.5 cm thick would cover the ground (assuming none sunk in or ran off—that's how much is actually falling through the air, some of which hits the baseball). The ball covers its trip quickly: 95 miles/h × 1,609 m/mile × 1 h/3,600 s = 42.46 m/s, meaning it covers the 18.44 m from the pitcher's mound in 0.43 s. At any given moment, it is obscuring (and therefore hitting or being hit by) raindrops over an area of πr^2 = 42.0 cm^2. Now in an hour, this much ground would accumulate 42 cm^2 × 0.5 cm = 21cm^3 of water. In 0.43 s, then, it will intercept 21 cm^3/h × 1 h/3,600 s × 0.43 s = 0.0025 cm^3, and since water has a density of 1.0 g/cm^3, the ball encounters 0.0025 g of water during the pitch. This is only 0.0025 g/145.2 g = 1.7×10^{-5} g of its original mass, or a 0.002 percent effect. We could get fancier by talking about where the water hits and the resultant momentum transfer, etc., but it is clear from getting this far that the effect is completely negligible.

5
INSIGHTS IN LINES AND DOTS

"Look at these two plots! ALL the red quasars are at low z—because they are so extincted we COULDN'T SEE THEM at higher z. It's the tip of a red iceberg!"[1]

Every morning I am confronted with at least seventy-five e-mails, a large percentage of which are either real-world spam, institutional spam, astronomical spam, or messages that indicate that I have to do something difficult or irritating, the latter being even less welcome than the spam. But the message above was from my long-time collaborator Dr. Rick White from the Hubble Space Telescope Science Institute, and his e-mails are always worth reading. The all caps and exclamations, however, were most unusual; Rick is from Tennessee (pronounced *tin-nuh-sea*, stretched out over about three seconds) and generally eschews exclamation points. I quickly opened the table of numbers (table 5.1) and the two plots (figure 5.1) he had attached.

This scientific discovery was completely buried in the mass of numbers in the table, but one glance at the graph and it was immediately apparent to me that the exclamation points were warranted.

Quasars are fascinating objects that generate the energy of a thousand trillion Suns in a space no larger than the Solar System. The engine that drives this enormous energy production is a supermassive black hole that has swallowed a billion stars and now lurks in the heart of a galaxy, ripping

TABLE 5.1 DATA ON THIRTY-FIVE QUASARS DISCOVERED BY OBTAINING OPTICAL SPECTRA OF ALL RADIO SOURCES WITH STELLAR COUNTERPARTS IN A PIECE OF SKY IMAGED WITH A VERY RED FILTER ON THE CAMERA

R.A. (J2000.0) (1)	Decl (J2000.0) (2)	Offset (arcsec) (3)	F_{pcore} (mJy) (4)	Size (arcsec) (5)	F_{int} (mJy) (6)	
10.01 30.555	+53.31 51.02	0.49	4.0	3.1 × 2.5	5.1	...
10.01 32.427	+51.29 54.52	2.06	1.4	4.4 × 0.0	1.8	
10.02 40.947	+51.12 45.36	0.83	3.8	< 2.5	...	
10.03 11.535	+50.57 05.39	1.44	1.6	~55T	24.3	
10.03 50.707	+52.53 52.41	13.08	6.9	~55T	113.0	
10.03 53.654	+51.54 57.29	0.50	1.1	6.3 × 0.0	1.5	
10.04 00.401	+54.16 16.96	0.83	9.0	< 2.5	...	

apart any star that wanders too close and sucking up its gas. The black hole itself is . . . well, black. But as the gas rushes toward its doom, it is heated to very high temperatures and glows brightly.

Quasars were discovered in the early 1960s as blue, star-like objects coincident with powerful sources of cosmic radio emission. At first, astronomers were extremely puzzled because an analysis of the quasar light showed signatures of elements apparently unlike any seen on Earth. Then, in one of those why-didn't-I-think-of-that-before moments, Professor Maarten Schmidt at the California Institute of Technology

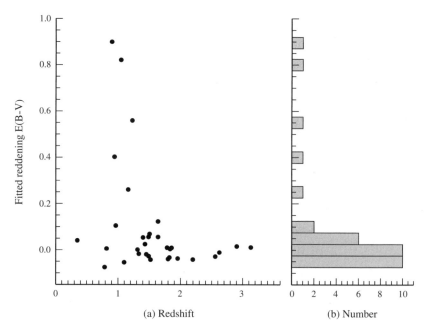

(a) Redshift   (b) Number

FIGURE 5.1 QUASAR REDDENING DISTRIBUTION

(a) A scatterplot (see pp. 98–104) showing the amount by which the quasar light has been made to appear redder as a result of the intervening dust versus redshift, or distance to the quasar. (b) A histogram (see pp. 97–100) of the number of quasars versus reddening ($E(B - V)$).

realized the light signature was in fact that of the most common element of all—hydrogen—but it was shifted to the red in such a way as to imply that the object was rushing away from Earth at 13 percent the speed of light. Quasars were not stars at all—they were the most distant objects yet observed, whose apparent flight from Earth was simply a result of the overall expansion of the universe. The enormous distance inferred from this first "redshift" (the "z" in Rick's note) implied that the quasar had to be intrinsically very luminous, outshining the combined light of the hundreds of billions of stars in its host galaxy. Today we have found quasars rushing away at up to 96 percent the speed of light, marking them as among the most distant objects seen. They provide us with a wealth of information about what the universe was like when it was less than 5 percent its current age.

As noted previously, the first quasars found were very blue in color, and most searches for additional such objects have focused on selecting the bluest objects from among the millions of star-like points of light that dot the nighttime sky. These searches have found many more blue quasars—the current catalog has more than 200,000 entries.

We had been working on our own search for four years—not because it was hard but because it was relatively boring. Finding a few dozen more quasars when 200,000 are known was just not the most exciting project we had underway. The one potentially interesting aspect of our search was that, rather than looking for blue objects, we were looking for red objects that distinguished themselves from normal stars by being bright sources of radio emission (just like Schmidt's original quasar). Our contrarian approach was designed to see if prior workers had missed out on a putative population of red quasars because of the blue bias in their selection technique (see chapter 9 on selection effects). In compiling the table of our findings, it was clear we had indeed found some very red quasars that appeared to have had their blue light filtered out by clouds of dust in their host galaxies (much as the Sun gets redder near sunset when its light passes through more and more of the Earth's atmosphere[2]).

We wrote up the results of our survey, including the data presented in table 5.1, and sent it off to the *Astrophysical Journal* to be reviewed.[3] After receiving the anonymous referee's comments, we began preparing a revised version of the article when Rick was inspired to make the plot reproduced in figure 5.1.

The discovery this new plot revealed was significant: we found not only some unusually red quasars but also that the five reddest quasars (the ones highest on the graph in figure 5.1) were the five most luminous objects—these previously overlooked red quasars were putting out more energy each second than the more "normal" bluer ones. In addition, the five reddest were among the quasars closest to us. This combination of properties meant that we had uncovered the tip of a red quasar iceberg—that forty years after quasars were discovered as distant blue objects we were just beginning to realize that astronomers had missed a huge portion of the population. The graph made this immediately apparent.

THE POWER AND LIMITATION OF GRAPHS

I'm not the only scientist for whom a graph can generate great excitement. I once saw 2,000 astronomers rise for a standing ovation when the graph shown in figure 5.2 flashed on the screen at a conference. This graph represents the first results from a satellite designed to take the temperature of the universe—literally. The Cosmic Background Explorer (COBE) had exquisitely sensitive instruments tuned to measure the faint afterglow radiation from the Big Bang. When the universe was born 13.8 billion years ago, it was a very hot place—billions of degrees. It has been expanding and cooling ever since, and the radiation from this initial fireball has now dropped to a temperature only 2.7 degrees above absolute zero. Since the universe was quite smooth and uniform at the time of the Big Bang, we can use our basic model of radiation to calculate how much energy we should receive at each wavelength in the infrared and radio bands of the spectrum.

Figure 5.2 shows the COBE measurements superposed on the predicted curve. The match is extraordinarily good. The fact that these

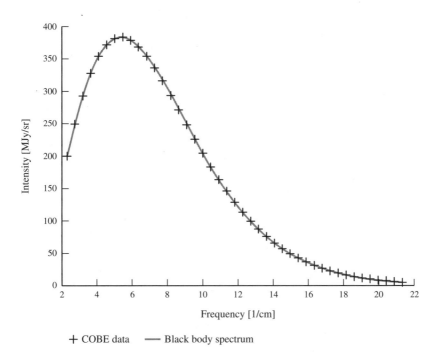

+ COBE data   —— Black body spectrum

FIGURE 5.2 COSMIC MICROWAVE BACKGROUND SPECTRUM FROM COBE

The spectrum (intensity vs. frequency) of the cosmic microwave background as observed by the Cosmic Background Explorer satellite. The solid curve is the theoretical prediction made 100 years ago; the crosses represent the satellite's measurements.

data allow us to measure the temperature today to an accuracy of one-thousandth of a degree (the latest measurements converge on a value of 2.725 ± 0.001 K) and the fact that our model for radiation, concocted to explain results in the laboratory over a hundred years ago, fit so beautifully the light from the universe's birth, provoked the ovation when the graph was shown at the annual American Astronomical Society meeting in 1994.

The notion that 2,000 astronomers and physicists would give a standing ovation to a graph is not prima facie evidence that we scientists represent an alien species. It is, rather, intimately linked to the evolution of our common species, *Homo sapiens*, and the complex circuitry in the eye and brain that has developed to assure our success in very different times.

SEEKING PATTERNS, PARSING DATA

A fundamental drive of all individuals and species is survival: to find enough food to eat and to avoid becoming food for someone else (and, for continuation of the species, to find a mate). The greater the distance from which one can detect tasty morsels and spot potential predators, the higher one's probability of survival. The vertebrate eye, connected to our specialized brain, is the remarkable device evolution has produced to allow long-distance observations. Over the past several hundred million years, the eye has been tuned to detect the most abundant wavelengths of light on Earth and to preprocess the huge amount of information coming in each second so the brain can keep up with the interpretations. One of the more valuable eye-brain processing systems allows for the rapid recognition of patterns.

If you wish to pass on your genes to successive generations, it has, through most of human history, been essential that you spot lions, leopards, tigers, and tarantulas before they spot you. The ability to quickly, indeed unconsciously, recognize patterns of spots and stripes is a valuable survival tool. The mammalian eye connects to well-adapted preprocessors in the retina and in the primary visual cortex that are especially sensitive to straight lines and edges in particular orientations; more sophisticated pattern recognition systems operate in other parts of the brain. These systems have been humming along for several hundred thousand years in their current form. Numbers, on the other hand, have only been part of human culture for 5,000 years, a relatively short timescale compared with our rate of evolution. It is unsurprising then that, confronted with the data shown in table 5.1, my colleagues and I failed to recognize the striking result present in our data, while within seconds of printing out the simple plots Rick had prepared, I was hammering out an excited e-mail discussing our new result.

This is not to suggest that numbers are not often the most effective way to present and digest some kinds of information. Representing the batting averages of the top ten players in the American League as a bar chart would not convey as much information, or convey it as crisply, as would a simple table of numbers. Plotting the proportion of ingredients in my secret recipe for chocolate mousse as a pie chart would likely keep it secret; it provides no clue as to the order in which the ingredients should be assembled or what intermediate steps must be taken by the mousse-maker.

Figure 5.3 tells you the batting averages of the top ten American League players, arranged in alphabetical order, at the end of the 2014 season. Think it looks a little ridiculous? That's probably because, considering the simplicity of the data, a bar chart is a superfluous way of displaying it. Just sticking to a table of numbers as in table 5.2 makes more sense.

The batting averages already represent partially digested results in that they are computed from measured data: number of hits/number of times at bat minus number of walks, number of times hit by pitch, interference calls, and sacrifices. Furthermore, there is usually no direct causal connection or interaction between the values achieved by different players (unless one happened to bat ahead of, or behind, David Ortiz in his prime). The bar chart of batting averages conveys no new information and reveals no emergent patterns.

TABLE 5.2 2014 AMERICAN LEAGUE BATTING LEADERS

Altuve	.341
Martinez	.335
Brantley	.327
Beltre	.324
Morneau	.319
Abreu	.317
Harrison	.315
Cano	.314
McCutchen	.314
Cabrera	.313

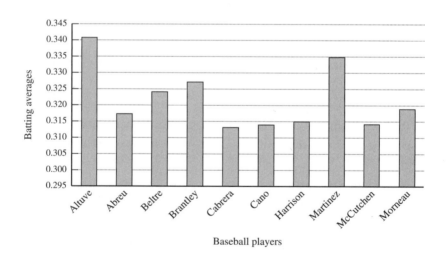

FIGURE 5.3 AMERICAN LEAGUE BATTING LEADERS FOR 2014

A histogram of the American League batting averages for the top ten players at the end of the 2014 regular season.

Likewise, the mousse recipe pie chart (figure 5.4) is simply a display of enumerated ingredients. Although there is undoubtedly some deep chemical reason why the proportions of chocolate, butter, bitter coffee, and sugar must be in the ratios specified for my mousse to achieve its intense flavor and velvety texture, the intermolecular interactions are unlikely to be illuminated by a pie chart, and simply combining the ingredients in the specified proportions in a food processor would lead to culinary disaster.

Whether 'tis nobler to graph or not thus requires judgment. But if the dataset is small and/or the numbers that represent measurements are in a computer with an easy-to-use graphing program, it can't hurt to plot

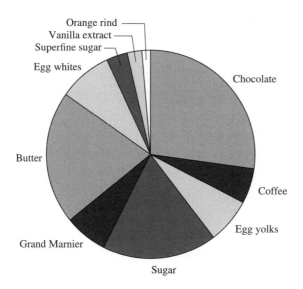

FIGURE 5.4 CHOCOLATE MOUSSE INGREDIENTS

A pie chart of my chocolate mousse ingredients by volume.

the data. Indeed, when faced with a large dataset in which each object of interest is characterized by several numbers, it is not uncommon for a scientist to just "plot everything against everything else"—to generate a large number of plots that represent the parameters of the dataset in different ways to help us look for the patterns we are so adept at seeing (see scatter plots later in this chapter). This apparently injudicious approach must be accompanied by a constant awareness that (1) we are so good at seeing patterns that we often see them where none exist (as in figure 5.5), and (2) the existence of a real pattern in a plot of variable A versus variable B does not necessarily signify that A controls B or that B controls A. Chapter 8 expands on this latter notion.

THE TYPES OF GRAPHS AND THEIR ESSENTIAL ELEMENTS

The word "graph" derives from the Greek *graphos*, which translates simply as "written." The most common current meaning, however, is the mathematical sense of a graph: a diagram showing the relation between variable quantities, typically of two variables, each measured along one of a pair of axes at right angles. As we shall see later in this chapter, we are not necessarily limited to two dimensions of information, and there are a number of graphical representations of data that involve more than two axes at right angles. The purpose of a good graph, however, is always to provide a visual display that summarizes accurately a large amount of information and that takes advantage of the eye-brain system's penchant for seeing patterns and inferring relationships. In the sections that follow we explore various types of graphs, distinguish between features that are arbitrary and those that are required, and draw attention to how graphs can propagate information as well as misinformation.

FIGURE 5.5 THE ORION CONSTELLATION

Seeing patterns in randomness: the three bright stars across the center of this sky image and the three vertically arranged ones lower down immediately catch the eye. Rather than forming the belt and sword of the hunter Orion, however, they are all at different distances and have no physical relationships whatsoever.

THE TIME SERIES

The raw materials for making a graph are data (observations or measurements of a physical, biological, or social system), or numbers (usually generated by a computer) that represent the predictions of a mathematical model (for more on models and data, see chapter 9). In most instances, these data are represented in numerical form—as pairs, triplets, or longer series of numbers that correspond to some measurable attributes of the system under study.

Take the Dow Jones Industrial Average (DJIA)—a widely followed single number reputed to be a measure of the strength of the U.S. economy. Each day in the *New York Times* and many other papers one can find a record of this average "minute by minute." Whether any profound meaning can be read into such data, there clearly is a lot of it: the DJIA is quoted to seven significant figures (of which only two actually contain economic significance) for each of the 390 minutes the stock exchange is open. It is clearly easier to represent this large collection of numbers in a graph. The display of any quantity plotted as *a function of*[4] time we call a time series plot. Each measurement of the DJIA comes with a paired number, the time of the measurement. The series of number pairs would look like table 5.3 or, in a time series plot, figure 5.6.

In most instances, time series plots use the *x*-axis as time and the *y*-axis as the quantity being measured. It is worth noting that this convention is, mathematically speaking, completely arbitrary. Furthermore, it is far from universal; e.g., some geological graphs have the time axis running from the present on the left and backward in time toward the right, whereas plots that represent evolutionary trees of life almost always have time running up the *y*-axis. The convention used here ties the stock market graph to our normal use of language. We talk about the DJIA going up (to higher numbers) and down (to lower numbers), not left and right, even though exactly the same information is displayed if we switch axes (figure 5.7).

TABLE 5.3 DOW JONES INDUSTRIAL AVERAGE ON FEBRUARY 17, 2015

Previous day's closing	18,018.55	12:40	18,019.36
		12:50	18,020.29
9:30	18,020.03	13:00	18,022.16
9:40	17,954.68	13:10	18,029.95
9:50	17,969.01	13:20	18,035.34
10:00	17,986.86	13:30	18,049.25
10:10	17,993.45	13:40	18,039.55
10:20	17,977.93	13:50	18,043.66
10:30	17,982.34	14:00	18,039.26
10:40	17,981.24	14:10	18,040.35
10:50	17,986.14	14:20	18,037.91
11:00	17,973.93	14:30	18,034.35
11:10	17,973.06	14:40	18,033.00
11:20	17,974.59	14:50	18,043.44
11:30	17,987.94	15:00	18,029.47
11:40	17,997.07	15:10	18,039.49
11:50	17,996.65	15:20	18,038.27
12:00	17,997.38	15:30	18,041.00
12:10	18,003.27	15:40	18,023.66
12:20	18,003.41	15:50	18,020.83
12:30	18,005.03	16:00	18,040.51

An enterprising young newspaper editor bent on creating a new style for his paper would thus be mathematically correct to present the stock market averages in the second form, although his tenure as editor might be short. Although the notion of the stock market increasing in value as it moves to the right might comport with political labels, our sense of being, pulled down by the gravity of the situation and soaring upward to heaven, all militate against such a change in the convention. A scientific example of the utility of time series plots is shown in box 5.1, where, after plotting

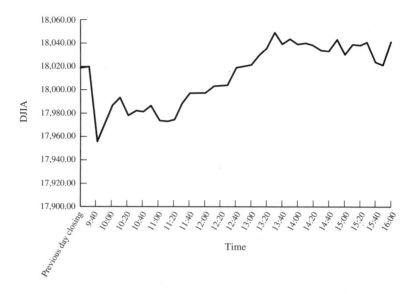

FIGURE 5.6 DOW JONES INDUSTRIAL AVERAGE

Minute-by-minute time series plot of the Dow Jones Industrial Average (DJIA) on February 17, 2015.

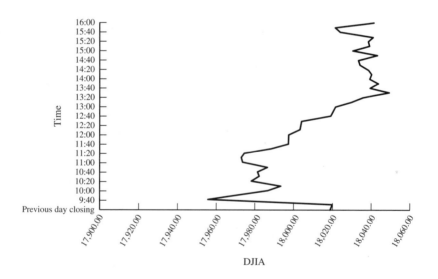

FIGURE 5.7 DOW JONES INDUSTRIAL AVERAGE REIMAGINED

The same Dow Jones Industrial Average (DJIA) graph from figure 5.6 simply rotated by 90 degrees. There is no reason other than convention that we could not make the plot this way—it conveys exactly the same information.

BOX 5.1 A SCIENTIFIC TIME SERIES PLOT EXAMPLE

We show in table 5.4 the record for the carbon dioxide concentration in the Earth's atmosphere as measured each month since January of 1958 at an observing station on Mauna Loa in Hawaii. This project was begun nearly sixty years ago by Charles Keeling and was subsequently carried on by his son. It is an iconic graph, representing the dawn of the Anthropocene.

The dense table of numbers makes it extremely difficult to extract any meaningful trends, but a time series plot (figure 5.8) reveals two striking patterns in the data: annual wiggles with a generally consistent amplitude of five to seven parts per million and a rapidly rising average value that indicates a net increase of eighty-six parts per million (or 27 percent) in the past fifty-six years.

Furthermore, laying a straightedge over the midpoints of the annual cycles shows that the slope, or rate of the increase, is increasing with time. The yearly wiggles arise from the extraction of CO_2 from the atmosphere by plant growth in the summer months, and its reintroduction to the atmosphere as decaying vegetation releases CO_2 in the winter and spring (see chapter 10). The long-term trend is a consequence of burning fossil fuels, a process that is releasing the CO_2 sequestered over 200 million years during of the age of the dinosaurs in a few centuries. The accelerating increase is a consequence of our ever-rising demand for energy.

the morass of numbers in the database, the human impact on the Earth's atmosphere is apparent. We will discuss this graph and its implications further in chapter 10.

AXES

In a well-constructed graph, each axis (and there can be more than two) will have a label and a series of numbers marking the length of the axes

TABLE 5.4 MONTHLY ATMOSPHERIC CARBON DIOXIDE MEASUREMENTS

	Jan	Feb	Mar	Apr	May	Jun	
1958			315.71	317.45	317.50		•••
1959	315.58	316.47	316.65	317.71	318.29	318.16	
1960	316.43	316.97	317.58	319.03	320.03	319.59	
1961	316.89	317.70	318.54	319.48	320.58	319.78	
1962	317.94	318.56	319.69	320.58	321.01	320.61	
1963	318.74	319.08	319.86	321.39	322.24	321.47	
1964	319.57				322.23	321.89	
1965	319.44	320.44	320.89	322.13	322.16	321.87	
1966	320.62	321.59	322.39	323.70	324.07	323.75	
1967	322.33	322.50	323.04	324.42	325.00	324.09	
1968	322.57	323.15	323.89	325.02	325.57	325.36	
1969	324.00	324.42	325.64	326.66	327.38	326.70	
1970	325.06	325.08	326.93	328.13	328.07	327.66	
1971	326.17	325.68	327.18	327.78	328.92	328.57	
1972	326.77	327.63	327.75	329.72	330.07	329.09	
1973	328.54	329.56	330.30	331.60	332.48	3332.07	
1974	329.35	330.71	331.48	332.65	333.09	332.25	
1975	330.40	331.41	332.04	333.31	333.96	333.59	
1976	331.74	332.56	333.50	334.58	334.57	334.34	
1977	332.92	333.42	334.705	336.07	336.74	336.27	
1978	334.97	335.39	336.64	337.76	338.01	337.89	
1979	336.23	336.76	337.96	338.89	339.47	339.29	

⋮

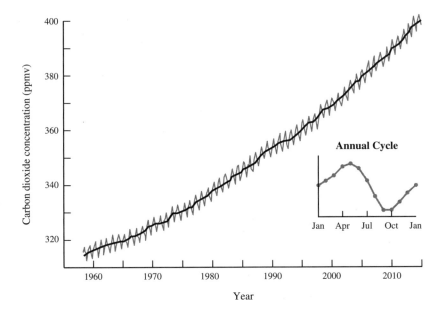

FIGURE 5.8 ATMOSPHERIC CARBON DIOXIDE MEASURED AT MAUNA LOA, HAWAII

A time series plot of atmospheric carbon dioxide concentration in the atmosphere (in parts per million by volume) measured at the Mauna Loa Observatory in Hawaii in a project begun by Charles Keeling. The inset shows an expanded version of the annual cycle caused by the seasons; the accelerating upward trend in the main plot is a consequence of fossil fuel burning (see chapter 10).

at fixed intervals. The labels should specify both what is being plotted (in a descriptive word, phrase, or abbreviation) along with the units (if any) employed. (DJIA is in an arbitrary system of "points," whereas time is either in minutes, days, or years depending on the timescale covered by the data.) Note that neither axis needs to start at zero. The tick marks and numbers that indicate the intervals can be spaced in either a linear or logarithmic fashion.

In some instances, both the left and right or the top and bottom of a plot will be utilized—with different scales and different labels. A simple example of this might be to provide alternative units so a single graph can serve multiple audiences. For instance, a car manufacturer might wish to display the fuel efficiency of a particular model as a function of driving speed (all cars have a peak efficiency of fuel use that falls off for either higher or lower speeds). For a European audience, this graph (figure 5.9) would be presented as the fuel used in kilometers per liter on the left side (y-axis) versus the speed in kilometers per hour on the bottom (x-axis); as such, it would leave the average American driver clueless. To avoid the cost of having to design and print a new

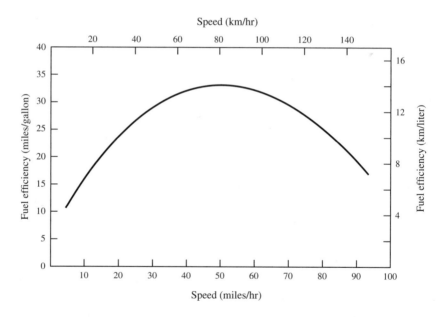

FIGURE 5.9 AUTO FUEL EFFICIENCY VS. SPEED

A plot of fuel efficiency as a function of driving speed, where two sets of alternative units are used on each axis.

graph for the American market, the manufacturer could simply print different scales and labels on the right-hand and top sides of the graph using miles per gallon and miles per hour, respectively. This might leave out the schizophrenic Brits who now think in miles per liter for fuel efficiency, but it would provide the required information for most of the world market.

It is also not uncommon to find more than one set of data on a single plot; in such a case the different axis labels refer to the different curves. For example, it has been known since Galileo first discovered them that sunspots—huge magnetic storms on the Sun that appear as dark blotches on its surface—vary dramatically in their frequency of occurrence. In addition to a regular eleven-year cycle of waxing and waning, changes occur on timescales of centuries and perhaps even longer. Much speculation has centered on whether these spots affect the Earth's climate. Figure 5.10 explores this possibility by including both the record of red sand grains dropped by icebergs in the North Atlantic and the rate at which radioactive carbon-14 (see chapter 10) is produced in the atmosphere, labeling the former on the left-hand vertical axis and the latter on the right. The extremely good correspondence between the two curves offers tantalizing evidence that increased solar activity—which wraps Earth in a protective magnetic blanket that screens the atmosphere from radioactivity-producing cosmic rays—is linked to significant increases in Earth's temperature (see the caption for figure 5.10 for a fuller explanation). When two seemingly unrelated measurements are correlated (see chapter 8), it suggests that they may both be caused by a third phenomenon—in this case, the changing energy output of the Sun. Plotting the two time series on the same graph (figure 5.10) makes the relationship apparent.

In summary, although the distribution of points and lines on a graph may visually convey an immediate sense of pattern, the axis labels hold the key to interpreting this pattern and the possible meaning that underlies it. So *read them*!

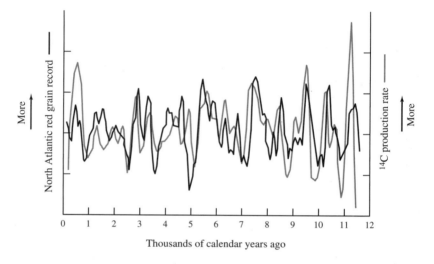

FIGURE 5.10 ICEBERG-DEPOSITED GRAINS AND C-14 PRODUCTION RATES

The concentration of red sand grains from Canadian soil (black curve and left *y*-axis) that appear in ocean sediment layers in the North Atlantic versus time from roughly the end of the last ice age to the present. Note that the time axis runs into the past from 0 (= today). The gray line (right *y*-axis label) shows the rate at which radioactive carbon-14 (C-14) is produced in the Earth's atmosphere. High-energy particles from the far reaches of the galaxy called cosmic rays slam into atmospheric Nitrogen atoms, transforming them into radioactive C-14. When the Sun is active, its magnetic field reaches out and deflects some of these rays from hitting the Earth, lowering C-14 production. Simultaneously, the more active Sun warms the Earth, causing icebergs to melt before the reach the North Atlantic and, thus, deliver fewer red grains. When solar activity wanes, the cosmic rays return to making C-14, and the cooler Earth allows icebergs from Canadian glaciers to reach more southerly latitudes where they eventually melt, dropping the embedded red soil grains to the ocean floor.

CAPTIONS AND LEGENDS

Sometimes just reading the axes is insufficient. Graphs such as figure 5.10 (and several of the others later in this chapter) also have legends. In figure 5.10, the axes tell you what is plotted, but you need to note the little vertical dashed and solid lines outside the box on each axis to know which curve refers to which axis; in this case, grain number is represented by the solid black line and carbon-14 production by the solid gray line. This graph also helpfully indicates the direction in which "more" of each quantity is found (note the graph does not provide the actual values for the two quantities—an unfortunate suppression of some of the data that must have been in hand to construct the plot).

This explanation of the symbols used in the graph is called a legend. Sometimes, as here, it is indicated simply; in more complex cases, legends are often included in a box inside the graph itself (see figures 5.27 and 5.33 for examples). Legends often contain key information necessary for the interpretation of the graph, so as with axes, read them.

In a well-crafted graph, the message should be clear from a cursory examination. However, for a full explanation of the data included and excluded, specification of the uncertainties involved, provenance of any models plotted in an attempt to explain the data, or simply to guide a full interpretation of the image presented, a caption is usually required. Even when a figure is clearly described in the text within which it is presented, it is worth repeating the salient points in a caption attached directly to the figure. Captions are often essential for a complete interpretation, so as with the axes and legends, read them.

THE BAR CHART

Even simpler than the time series is a frequently used graph known as the bar chart. Like all graphs, the bar chart is simply a way to collect a large number of data points and represent their content at a glance. For example, if we were to send ten teams of students out to measure the numbers of plants of various species found at several locations in the city, the basic datum each team would return with from each location is a single integer—the number of species found. Furthermore, these numbers are likely to have a limited range: there certainly cannot be fewer than zero species found, and our ecologist colleagues can tell us the number in any one location is unlikely to exceed twenty. The dataset from this kind of experiment can be well represented as a bar chart. Thus, we go from a table of numbers (table 5.5) to a simple bar chart (figure 5.11).

TABLE 5.5 NUMBER OF EACH SPECIES FOUND AT EACH SITE

Species Name	Site 1	Site 2	Site 3	Site 4	Site 5	Site 6	Site 7	···
Japanese Barberry	1	3	5	3	1	2	0	
Witch Hazel	0	1	0	2	0	2	0	
Holly	0	3	4	0	0	1	2	
Sheep Laurel	2	0	6	0	3	4	0	
Mountain Laurel	0	0	0	3	2	1	0	
Spicebush	3	5	4	1	4	2	1	
Pink Azalea	0	1	2	8	0	0	0	
Rose	7	5	8	10	0	1	0	
Common Blackberry	3	2	5	0	2	2	1	
Prickly Dewberry	0	1	3	3	0	0	1	
Bristly Dewberry	10	5	12	0	4	3	2	
Flowering Raspberry	4	2	0	1	2	3	0	
Late Low Blueberry	2	0	3	0	1	1	0	
Common Highbrush	0	1	3	2	3	2	1	
Blueberry	0	1	2	0	2	2	0	

⋮

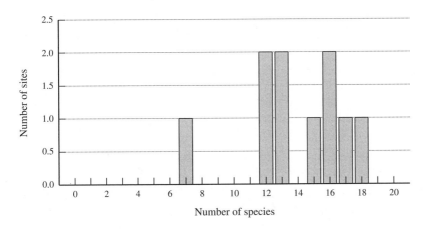

FIGURE 5.11 SPECIES DIVERSITY

A bar graph showing the number of sites versus the number of species found at each site.

The chart in figure 5.11 immediately shows that none of the locations chosen was completely devoid of plant life (no zeros) and that highly diverse sites were rare. The typical site had twelve to sixteen species, but the range extended from seven to eighteen. Comparing this chart to various mathematical functions, such as the statistical descriptions discussed in chapter 8, we can characterize the entire dataset with a few numbers such as the mean, range, and standard deviation. This compact description is then amenable to comparison with theoretical models for why plant diversity is distributed as it is.

THE HISTOGRAM

If our data consist of a set of measurements of some quantity that varies continuously, it is usually appropriate to create a modified form of the bar chart called a histogram. Suppose (as I did as a geeky youth) that I measured the temperature every day for two years at precisely 6:30 A.M. (the time I had to get up to catch the school bus). Trying to be a good young scientist, I was careful to measure the value to the nearest half degree (oblivious to the fact that this was almost certainly meaningless precision, given that wind speed, humidity, and the recent rate of temperature change—all of which affect the measurement—were not also recorded).

At the end of my observations, I had 731 data points[5] ranging from 17.5°F to 99.0°F. Since there were 164 possible values[6] in this range for me to record, there are at most a few values falling at each bar's location; some bars have zero entries, and only two had more than twelve. With this large number of possible bars and the relatively small number of measurements, a bar chart displays almost too much information and obscures general trends (see figure 5.12). It also misrepresents the physical quantity I was measuring to some degree (pun intended), implying that temperatures could only have values of 50.0 or 50.5, for example, when 50.25 or 50.1897 are equally probable. To acknowledge the approximate nature of my measurements and to produce a more informative graph, I need simply to collect the data into "bins" and plot them as a histogram. That is, I add up the number of measurements between 90 and 100, the number between 80 and 90, etc., and reduce the 731 numbers to 9. I then draw them as in the bar chart but with the bar edges removed, indicating that the measurements are continuous and that there is no sharp break between one bin and the next. Again, standard statistical measures of the resulting histogram (figure 5.13) allow an even more compact description of my observations to help facilitate comparisons with models.

THE SCATTER PLOT

One of the most common circumstances in the lab, in the field, or at the telescope is that one collects measurements resulting in two or more

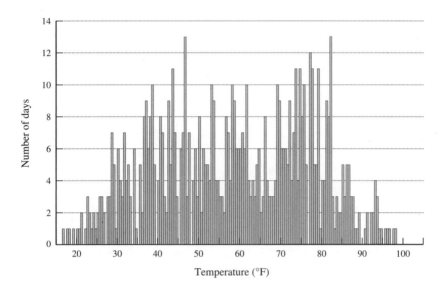

FIGURE 5.12 DAVID'S TEMPERATURE RECORDS

Bar chart of my temperature data showing the number of days over two years on which each temperature was recorded.

numbers that describe each object of interest, be they atoms, deer ticks, stars, or a collection of academics.

A recent article on the fraction of women in a wide variety of academic disciplines offers an excellent example of the use of the scatter plot (and the correlation analysis that follows from it—see chapter 8 for more on this). Leslie et al.[7] conducted a large survey filled out by 1,820 faculty, postdoctoral fellows (recent PhDs pursuing further research training), and graduate students in thirty different academic disciplines at highly ranked public and private research universities across the United States. They collected data on the number of hours worked each week on campus and off and the selectivity of each field as measured by the fraction of graduate applicants accepted into their PhD programs, as well

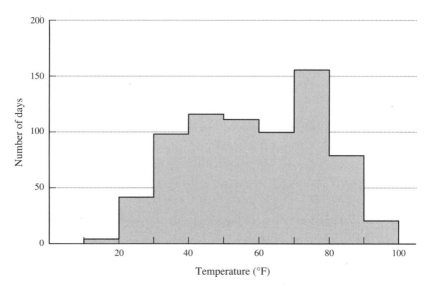

FIGURE 5.13 DAVID'S TEMPERATURE RECORDS REBINNED

A histogram of the same data presented in figure 5.12 showing the broad trends more readily.

as information about attitudes on a variety of issues such as the degree to which a field requires systematizing information versus empathizing with others, and, to test the authors' specific hypothesis, the degree to which special innate talent is required for success in the field. For this last datum, respondents were asked the extent to which they agreed with the following statement: "Being a top scholar of [insert discipline] requires a special aptitude that just can't be taught." They were also asked what they thought the prevailing view in their field was on this question, independent of their own views.

This survey produced a lot of data—the answers to fifteen questions by more than 1,000 people. In addition, the authors collected the percentage of PhDs awarded to women in 2011 in the United States for each of the thirty disciplines studied; this ranged from 76 percent in art history to approximately 18 percent in computer science and physics and 16 percent in music composition. As noted previously, faced with a lot of data, a simple first step is to plot everything relative to everything else. Some of the graphs from this dataset are shown in figures 5.14–5.17.

In figure 5.14, the points are randomly scattered, and no trend is apparent—there is no apparent relationship between how much academics work at their jobs and the fraction of women in the field. Figure 5.15, which includes only hours worked while on campus, shows a slight trend in the sense that the more hours worked on campus, the fewer women there are in the field. Whether this trend is significant, however, must await the quantitative analysis we explore in chapter 8.

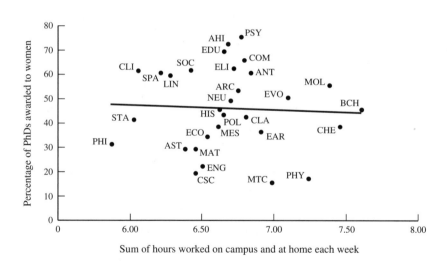

FIGURE 5.14 WOMEN PHDS VS. HOURS WORKED (10-HR INCREMENTS)

A scatter plot showing the fraction of PhDs awarded to women in a variety of academic fields as a function of the sum of hours worked on campus and hours worked at home each week by practitioners in each field. (*Source:* Leslie et al. 2015.)

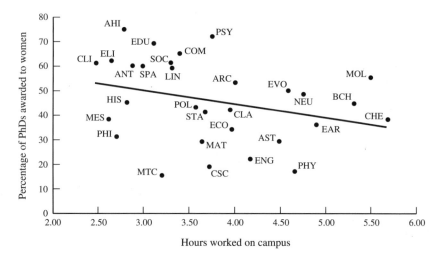

FIGURE 5.15 WOMEN PHDS VS. HOURS WORKED ON CAMPUS (10-HR INCREMENTS)

A scatter plot showing the fraction of PhDs awarded to women as a function of the number of hours worked on campus. (*Source:* Leslie et al. 2015.)

In figure 5.16, the trends are quite striking. In both the STEM (science, technology, engineering, and mathematics) fields and the social science/humanities disciplines, the greater the belief in the importance of innate ability, the smaller the fraction of women who obtain PhDs in that field. The authors conclude this is by far the most important effect among the various explanations they explore for gender differences among different fields.

Figure 5.17 shows an even stronger relationship (in chapter 8 we'll call it a tighter correlation) between the self-reported attitudes of individuals in each field relative to their assessment of the prevailing attitudes in their respective disciplines. Although perhaps unsurprising, it does suggest a slightly disturbing conformity of views.

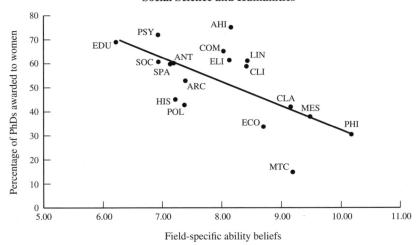

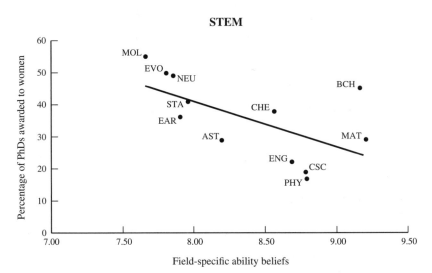

FIGURE 5.16 WOMEN PHDS VS. "FIELD-SPECIFIC ABILITY BELIEFS"

Scatter plots of the fraction of PhDs awarded to women as a function of "field-specific ability beliefs," where a higher number means there is a stronger belief that special, innate, unteachable talent is required to succeed in the field. The two plots represent these distributions for the twelve science, technology, engineering, and math (STEM) fields and the eighteen social science/humanities fields studied. (*Source:* Leslie et al. 2015.)

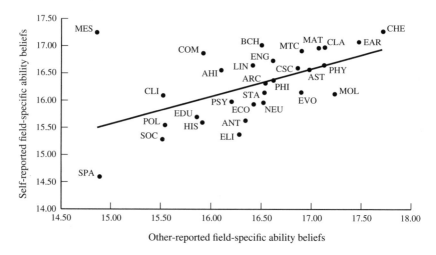

FIGURE 5.17 "FIELD-SPECIFIC ABILITY BELIEFS": SELF-REPORTS VS. ASSESSMENT OF PRACTITIONER VIEWS

A scatter plot of the self-reported field-specific ability beliefs versus the respondent's assessment of such beliefs in his or her field. (*Source:* Leslie et al. 2015.)

Many more plots could be made from the data Leslie et al. collected, and in their article they systematically examine the relationships among many of their variables and perform various tests to look for biases in their data. They could not answer the question as to whether it is true that special, innate ability is required for success in some fields, but they demonstrated quite convincingly their hypothesis that the strength of a belief in this requirement for a given discipline correlates with the fraction of PhDs awarded to women in that discipline. Whether this belief *causes* the underrepresentation of women in some fields is another matter that we will take up in chapter 8.

ADDING DIMENSIONS: THE CONTOUR PLOT

While the sheet of paper on which a graph is drawn is a two-dimensional (2-D) surface, our data collection need not be limited to only two numbers representing each object of interest. When three or more numbers make up each data point, it is necessary to develop additional dimensions for our graphical representations. Perhaps the most familiar of these is the contour plot.

If you are about to set off on a hike in unfamiliar territory, it is unwise to use a standard 2-D map to estimate how long it will take to get from point A to B. A map does, of course, come with a scale that you can use to convert inches to miles (or, preferably, centimeters to kilometers) and thus determine the "distance" between the two points. But walking the twelve miles from the Broadway Bridge (A) to Battery Park (B) in New York City and the twelve miles from Mountain Road in Cascade, Colorado (A) to Route 67 (B) are very different experiences (see figures 5.18 and 5.19, respectively). The surface of the Earth is three-dimensional, and the number of steps you take (and the number of Power Bars required) will depend not only on the 2-D map distance but also on the changes in elevation you must traverse. To provide this third dimension, we often use a contour plot.

Each point along the route can be represented by its position in a 3-D space. Your origin is at a particular longitude, latitude, and elevation above sea level; your endpoint likewise requires three coordinates to represent its position, as does every point in between. In order to display this third dimension (elevation) on the 2-D surface, we draw lines that connect nearby points with the same value in that parameter; e.g., all points 100 meters above sea level will be connected with a line until it either makes a continuous closed loop or gets to the edge of the page. Points 200 meters high are then linked by a separate line, etc. Figures 5.20 and 5.21 show the respective results for the two maps described in the previous paragraph.

FIGURE 5.18 STREET MAP OF MANHATTAN AND SURROUNDING AREA

The scale is 1 cm = 1 km. The point marked A is the Broadway Bridge to the Bronx, and B is Battery Park, at the southern tip of the island, a little over 12 miles (20 km) away.

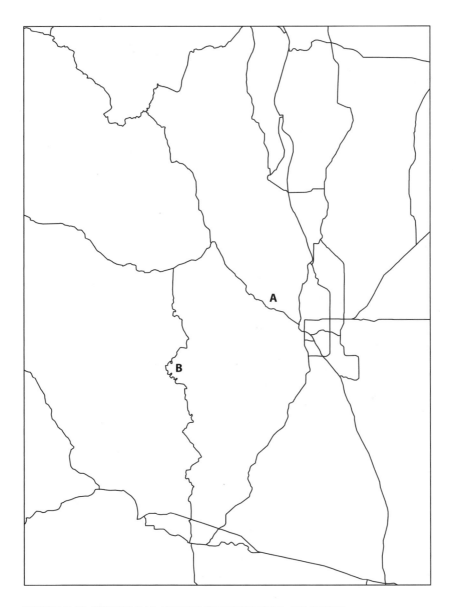

FIGURE 5.19 STREET MAP AROUND INDEPENDENCE, COLORADO

The scale is 1 cm = 1 km, but the road density in this neighborhood is clearly much lower than in Manhattan. The point marked A is Mountain Road in Cascade, Colorado, and point B is on Route 67; they are the same 12 miles (20 km) apart as points A and B in figure 5.18.

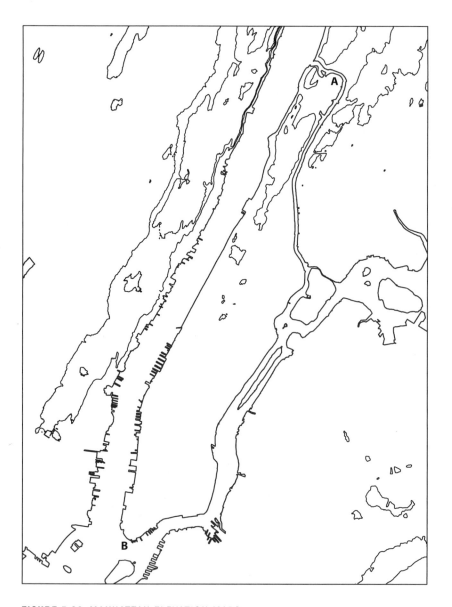

FIGURE 5.20 MANHATTAN ELEVATION MAPS

Contour map of Manhattan (with contours every forty meters). Points A and B are the same as in figure 5.18.

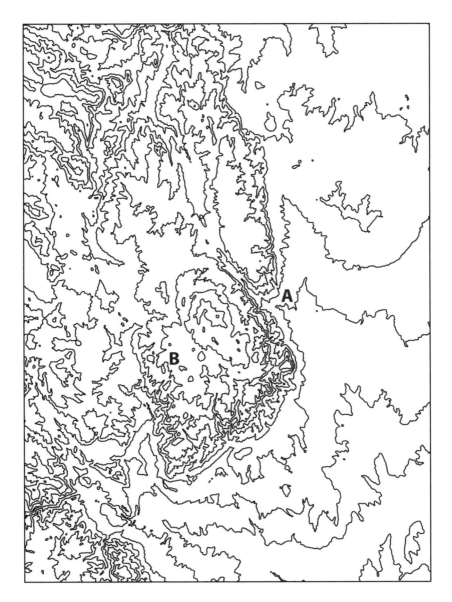

FIGURE 5.21 COLORADO ELEVATION MAP

Contour map of Pike's Peak in Colorado (with contours every 400 meters, ten times greater than in figure 5.20). Points A and B are the same as in figure 5.19.

Clearly the two maps reveal a very different picture of how your day will unfold. In Manhattan, the trek will be nearly two-dimensional; only Washington Heights at the north end of Manhattan and Morningside Heights below it represent some steps expended going up and down instead of forward. In Colorado, however, the route charted—while the same number of inches on the map or miles across the surface of the globe—will require the ascent and descent of 14,114-ft. Pike's Peak from a starting altitude of 7,000 ft.—that's over one and a third miles up and over half a mile down, or almost two extra miles (assuming you could actually walk in a line as straight as that of the Manhattan street grid).

Contour lines provide a vivid representation of the third dimension once you train your brain to recognize what they mean. Elevation lines squeezed tightly together mean that the elevation is changing rapidly (translation: a very steep route to be avoided). Jagged lines mean frequent changes in elevation (up and down, up and down); paths along which elevation line values (typically labeled at intervals along each line) constantly decrease is where you get to relax (going down hill)—unless they are very close together, in which case falling down hill may be the uncomfortable result.

Contour lines need not, of course, only represent elevation on a map. They can be used to illustrate the third dimension of any dataset in which each point is described by three measurements, e.g., the radio wave brightness of an exploded star's remnant on the sky (figure 5.22).

Maps and images are not the only graphical displays in which extra dimensions are required to illustrate the full suite of data available. It is often useful to display the attributes of different types of objects on the same plot using different symbols—dots, triangles, stars, etc. For example, in following up our apparent discovery that lots of hidden, red quasars were lurking in the universe, we set out to uncover ways of finding them efficiently. First, we went even redder than red by looking at infrared images of the sky to find candidates. We then followed up each candidate by taking its spectrum—breaking up the light into its constituent wavelengths, much as a prism breaks white light into the colors

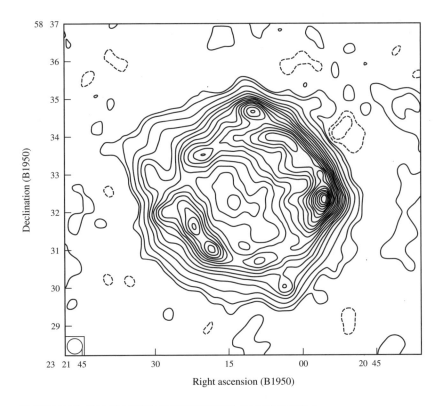

FIGURE 5.22 RADIO SURFACE BRIGHTNESS OF CASSIOPEIA

A radio image of the sky at the location of a titanic stellar explosion that blew apart a massive star in the year 1665. The *x*- and *y*-axes represent position on the sky (analogous to longitude and latitude on the Earth's surface), and the contour lines represent radio brightness; the "peaks" here are more than 100 times brighter than the map average. The inset in the lower left corner shows the resolution of the telescope's beam.

of the rainbow. Analyzing these spectra can tell us an object's distance and whether it is a star, a normal galaxy, or a quasar. We displayed the results in figure 5.23.[8]

The axes represent the quasar "colors" calculated by combining visible and infrared measurements. The letters J, K, and R represent standard wavelength

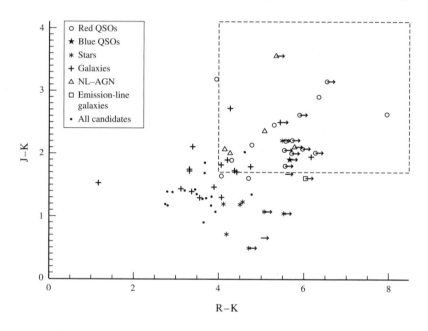

FIGURE 5.23 COLOR-COLOR PLOT FOR QUASAR CANDIDATES

Infrared (J–K) versus optical (R–K) colors for quasar candidates in our survey. The results of observations designed to yield distances and source classifications are illustrated by the different symbols displayed. The box enclosed by the dashed lines in the top right identifies a region in which 50 percent of the candidates turn out to be our quarry: red quasars. (*Source*: E. Glikman, M. D. Gregg, M. Lacy, D.J. Helfand, R.H. Becker, & R.L. White. 2004. "FIRST-2Mass Sources Below the APM Detection Threshold: A Population of Highly Reddened Quasars." *Astrophysical Journal* 607: 60–75.)

bands in which astronomers observe. The difference between pairs of bands tells us whether an object is redder or bluer (larger numbers indicate redder). The different symbols represent stars (*), galaxies (+), red quasars (o), and other denizens of the celestial zoo. Examination of the distribution of different symbols in the plot allows us to define a region (the upper right-hand box) in which we have a 50 percent chance of finding a red quasar; outside this region, most of the candidates turn out to be boring stars or galaxies.

LIMITS AND UNCERTAINTIES

The data we collect and wish to represent on a graph are rarely both complete and infinitely precise. Most measurements include a range of uncertainty or error (see chapter 7), and it is important to represent this on our graphs. Furthermore, we are often unable to obtain all the measurements we want, and this incompleteness is also essential to display.

There are standard conventions for representing both "errors" (uncertainties) and "limits" on graphs. Each point should have an associated uncertainty in the measured value of the quantity of interest. In this astronomical "spectrum," we plot the amount of radiation the source is emitting relative to the frequency of light (in this case, radio and infrared waves) emitted (figure 5.24).

In this representation, the dot represents the measured value, and the vertical length of each "error bar" illustrates the degree of uncertainty for the point through which it passes. It is important to specify the convention used in drawing the length of the bar. It can represent the entire range of measurements or some other indication of uncertainty (the options will be explored in chapters 7 and 8). The choice should be clearly spelled out in a figure caption (an admonition almost universally ignored in newspapers and all too infrequently followed by scientists).

For a scatter plot, measurements can have uncertainties in both coordinates. In the case of figure 5.25, for example, both vertical and horizontal

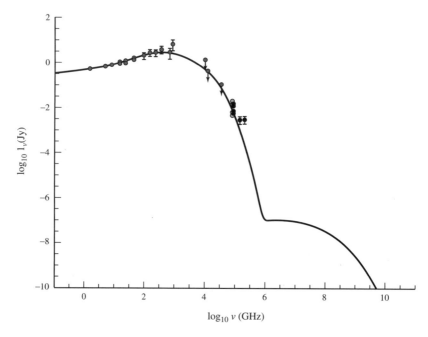

FIGURE 5.24 BROAD-BAND SPECTRUM OF THE MILKY WAY'S CENTRAL BLACK HOLE

The radio/infrared spectrum of the 3.6 million solar mass black hole at the center of the Milky Way. The error bars represent one-sigma uncertainties (chapter 7) in the measurements; where not shown, they are smaller than the points. Note that the frequencies at which we take the data are so precisely specified that the uncertainty is much smaller than the width of the dots, and no horizontal error bar is necessary. The continuous line drawn through the points is a theoretical model of what such a black hole should emit. Note that the point near 3.0 on the *x*-axis lies above the line, but only by about 1.5 times the length of its error bar; statistically speaking (see chapter 7), we should expect roughly one such deviation out of every ten measurements, so this point is *not* inconsistent with the model. However, the two points farthest to the right are many sigma above the predicted curve and thus indicate a problem with the model.

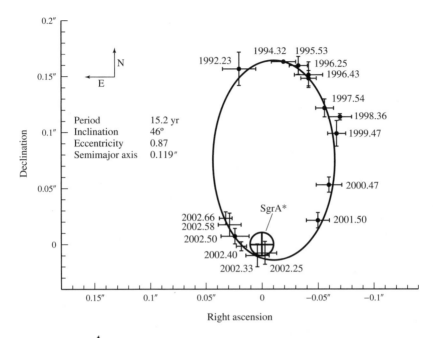

FIGURE 5.25 STELLAR ORBITS AROUND THE MILKY WAY'S CENTRAL BLACK HOLE

The positions as a function of time of a star orbiting the massive black hole at the center of our galaxy. Each point is labeled with the date of the observation and the uncertainties in the two coordinates on the sky (right ascension and declination, the equivalent of longitude and latitude on Earth) are indicated by horizontal and vertical error bars, respectively. SgrA* is the name of the central black hole; its position is indicated by a circle enclosing a cross, which represents its positional uncertainty. Determining the parameters of this star's orbit (indicated on the left) led to an accurate measurement of this black hole's mass. (*Source*: http://astronomycafe.net/qadir/BackTo266.html. Graphic by Rainer Schödel, Infrared and Submillimeter Astronomy Group at the Max-Planck-Institut für extraterrestrische Physik) error bars can be plotted on each point. If the number of points is large, and the hundreds of overlapping error bars would clutter the plot, it is acceptable to illustrate the "typical" error bars for a few points to give the viewer an indication of the uncertainties involved.

In a bar chart or histogram, each bin can be assigned a vertical error bar. This is usually based on the number of data points collected in each bin. Since the relevant statistics for most counting experiments follow the binomial distribution (see chapter 8), the error bars are most often displayed as representing some confidence interval from that distribution (e.g., given the measured value, there is a 95 percent chance that the true value lies between the upper and lower limit of the bar's length).

Data incompleteness can take several forms. A star may simply be too faint for our instruments to detect, or several patients in a large clinical study may disappear on a sailing trip around the world, preventing us from obtaining timely blood samples. In the case of the star, our data are not useless. They may not tell us the brightness of the star, but if we know the brightness of the faintest star our telescope could see, the nonmeasurement assures us that the star is fainter than this value. This is known as an "upper limit," and it is important to display such limits on our graph to avoid misrepresenting the results of our experiment. In the case of the missing patients, their absence may lead to a "lower limit" being indicated on our graphical display of the trial: "at least forty-three patients responded positively to the medication" (the forty-three measured plus, perhaps, some of the five whose measurements are missing).

Upper and lower limits are typically represented by arrows pointing in the appropriate direction: down or to the left for an upper limit, up or to the right for a lower limit (assuming the axes increase from the origin upwards and to the right—remember to read the axes carefully). In a histogram or bar chart, little arrows inserted into the bars themselves indicate how many of the measurements displayed in each bin are actually lower or upper limits. The black hole spectrum, shown in figure 5.24, includes several upper limits near the x-axis value of 4.0, all of which are barely consistent with the model prediction. For another example of a graph with limits, see figure 5.23.

TRICKS AND MISREPRESENTATIONS

As chapter 7 notes, statistics have a bad reputation; their misuse, deliberate or otherwise, can easily mislead the innumerate and confound all but the most careful scientist. Graphs have a less scurrilous reputation, but they too are subject to simple manipulations that can convey false or misleading impressions.

The simplest trick, often found in newspaper articles, is the suppressed zero. Consider two representations (figures 5.26 and 5.27) of the DJIA time series plots. The graph shown in figure 5.26 could be accompanied by the headline "Stocks Plunge on Interest Rate Worries." The precipitous decline in the curve around 3 P.M. looks ominous indeed. Would the plot shown in figure 5.27 carry the same headline? Could any reasonable person call the nearly invisible inflection near 3 P.M. a "plunge"? Probably not, leaving the headline writer bereft of emotionally charged words with which to describe the day's "market action." These two figures, of course, present exactly the same data. Which is more "effective"? Which is more accurate?

Another behind-the-scenes sort of manipulation to which one must always be alert applies primarily to bar charts and histograms: judicious binning. The person making the graph controls the widths and starting points of histogram bins. By trying many different combinations, purely random statistical fluctuations (chapter 6) can be made to look like significant results. The simplest, most straightforward choices are usually best: starting at zero, using integer bin intervals and equal bin sizes. (For an excellent, more thorough discussion on graphical methods and tricks, you may want to take a look at Edward Tufte's publications.[9])

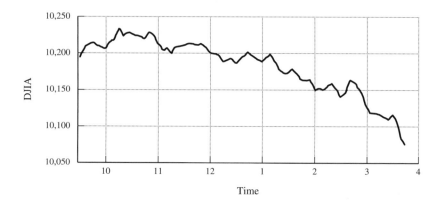

FIGURE 5.26 DOW JONES INDUSTRIAL AVERAGE WITH SUPPRESSED ZERO

The typical newspaper display of the minute-by-minute Dow Jones Industrial Average (DJIA); the impression is clearly of a substantial decline toward the end of the trading session. Note, however, that the origin on the *y*-axis is not zero or even close to zero.

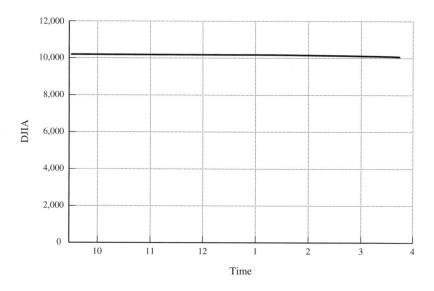

FIGURE 5.27 DOW JONES INDUSTRIAL AVERAGE WITH ORIGIN AT ZERO

The same graph as in figure 5.26 with *y* = 0 as the origin. The huge decline represented in the suppressed zero graph in figure 5.26 is now inconspicuously smaller than the thickness of the line representing the average.

There are occasions, however, when it is appropriate to experiment a little or to choose sizes that otherwise might not be optimal. For example, the two graphs shown respectively in figures 5.28 and 5.29 represent the same distribution of redshifts (= distances) for cosmic x-ray sources in the deepest map of the x-ray universe ever made (Chandra Deep Field-North[10]); figure 5.28 shows a fairly flat distribution between redshifts 0 and 1, whereas figure 5.29 shows dramatic

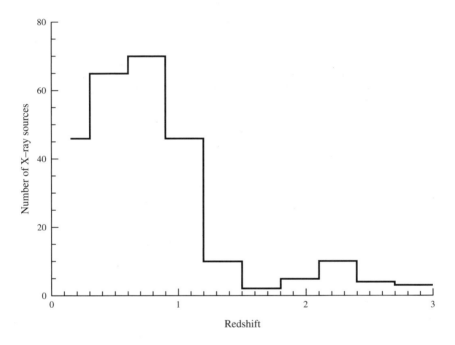

FIGURE 5.28 REDSHIFT HISTOGRAM FOR THE CHANDRA DEEP FIELD

A histogram of the number of *x*-ray sources versus their redshifts that correspond directly to their distances from Earth. We use a reasonable binning scheme starting at zero and use equal bin widths of 0.3 in redshift. It is clear that most of the objects have redshifts less than or about one.

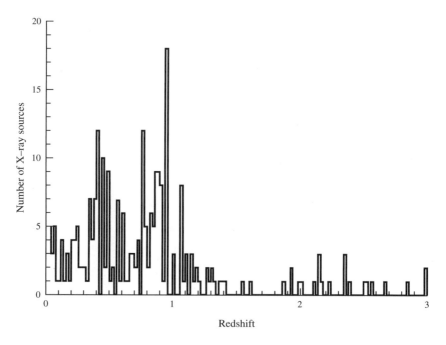

FIGURE 5.29 CHANDRA REDSHIFT HISTOGRAM REBINNED

The same data plotted with a bin width fifteen times smaller. Note the large narrow spikes, which represent physically clustered galaxies all at the same distance.

spikes at 0.4, 0.75, and, particularly, 0.95, where eighteen objects are found at the same distance (most of the rest of these narrow bins are underpopulated). The spikes almost certainly represent a real effect—x-ray galaxies clustering together at specific distances from Earth. Thus, figure 5.29 conveys additional information: not only are most x-ray galaxies found between redshifts 0 and 1 but they clump together in huge clusters as well.

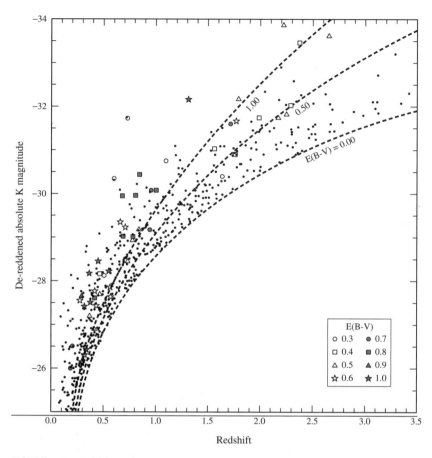

FIGURE 5.30 DISCOVERY OF A POPULATION OF RED QUASARS

Infrared luminosity (energy emitted each second) versus redshift (distance) for quasars in our radio survey sample. Small dots represent normal blue quasars discovered using standard techniques. Large symbols represent red quasars discovered by our program using infrared selection that employs the graph shown in figure 5.23. The legend appears in the bottom left; increasing $E(B - V)$ values indicate increasing amounts of dust obscuration. The dotted lines indicate the limits of our survey for various amounts of obscuration; e.g., no quasar with a reddening $E(B - V)$ of 0.5 or greater can fall below the dashed line labeled 0.50 (note that none do, since the black dots all have $E(B - V) \sim 0$ by dint of the way they were discovered—as blue quasars). Note that the most highly reddened quasars are the most luminous (we could not see them unless they were), and only moderately reddened (unfilled symbols) quasars are visible at large distances (again, they would be too faint if they were heavily obscured).

SO HOW MANY RED QUASARS ARE THERE ANYWAY?

To summarize many of the ideas presented in this chapter, I point you to figure 5.30, which is from a paper in our now long-running series on the discovery of red quasars.[11] Prepared by my former graduate student Dr. Eilat Glikman, who is now a professor at Middlebury College, it shows the distribution of the dozens of red quasars we found, along with the previously known population. The axis labels, multiple symbols, lines, and shadings tell a story about enormous black holes lunching on stars in the privacy of their enshrouding dust. It tells their ages, distances, and the sizes of their domains. A very large table of numbers, representing hundreds of hours of work and hundreds of thousands of dollars worth of telescope time, are suc-cinctly summarized in this plot, which also points the way to the next step in our research. In science, a graph is often worth more than a thousand words.

INTERLUDE 2

Logic and Language

While celebrating the explanatory—and revelatory—power of graphs as an essential feature of the scientific perspective, as well as the importance of numbers in a rational description of Nature, it is appropriate to acknowledge that the primary means of communication among scientists, and between scientists and the larger world, remains words. And words, in turn, are an essential component of logical thinking. Unfortunately, in addition to rampant innumeracy, ignorance of very basic physical concepts, the neglect of logical reasoning, and the sloppy use of language also contribute significantly to the glut of nonsense with which we are bombarded in the Misinformation Age. This interlude provides a modest sampling of the problems. Take this clip from a March 28, 2014, *New York Times* article: "Globally, sea levels have risen eight to 10 inches since 1880, but several studies show that trend accelerating as Arctic sea ice melts."[1]

Let me propose an experiment. Fill a glass of water nearly to the top and then add a couple of ice cubes until the water level is even with the top of the glass. Let it stand for an hour or so until the ice cubes have melted. Did the glass overflow? Of course not. Ice is less dense than water.

That's why, initially, the ice cubes were floating, sticking above the rim of the glass. As they melted, the water molecules they contained snuggled closer together, taking up just enough less space to allow the water in the parts of the cubes sticking over the rim to settle in. Archimedes understood this 2,500 years ago, but the basic logic of his conclusions about density and displacement seems not to have gained general distribution.

Melting Arctic sea ice has *no* direct effect on sea level. If all 10,000 km³ of the sea ice melted (as is now predicted during the Arctic summers in just a few decades), the sea level would not rise 1 mm. Such an event *will*, eventually, contribute to global warming, however, because sea ice reflects much more of the Sun's light back into space than open water, and the more of the Sun's energy that is absorbed, the higher the temperature. That could accelerate melting of ice on land, which does contribute to sea level rise, although contrary to popular conceptions, the increasing sea levels we are experiencing today primarily result from the thermal expansion of the oceans as temperatures rise (warm water takes up more space), not the additional fresh water running into the sea from melting glaciers (see chapter 10).

There are numerous examples of simple physical principles such as this that are violated by "common wisdom." When bicycling in Central Park, my spouse has repeatedly insisted that I coast down hills faster than she does because I weigh more. Aristotelian thinking (heavy things fall faster than light ones) is not the answer—I have a better bike.

I remember my father, a farmer, solemnly intoning that winter nights were colder when the Moon was full. Since the Moon has zero impact on Earth's weather, I knew this had to be false, but it took me decades to realize where this idea came from. Recently, it dawned on me. On clear winter nights, radiational cooling does make it colder (no clouds to trap any of the heat radiation escaping into space), and, of course, one can only see the full Moon when it is clear. On nights with thick clouds, it is warmer, but one never sees the full Moon then. So what my father had observed (or more likely, what had been passed down to him by generations of farmers in his family) was that it was colder on nights with a full Moon. It's a real correlation—containing no information whatsoever about the Moon's effects on Earth.

And while we are skewering widely held myths, we do not have seasons because we are closer to the Sun in summer; in fact, we are farthest from the Sun during the first week of July. Seasons arise because of the tilt of the Earth's axis with respect to the plane of its orbit around the Sun. This leads to a change in the angle of the incoming Sun's rays over the course of the year. A steeper angle (in winter) means each square meter of Earth absorbs less energy and thus is colder. The common view again marks a failure of logic. Many people know that the Southern Hemisphere has seasons opposite to those we experience in the Northern Hemisphere. If that's true (and it is), how could it possibly be the distance of the earth from the Sun that causes seasons?

WORDS

Having established myself as a pedant, you probably won't be too surprised that my penchant for the precise use of numbers extends to the precise use of the English language. English is rich in words, and each of them has one or more distinct definitions and appropriate places for use. Billion and million mean very different things and should be used for what they actually stand for, not interchangeably with "some big number, whatever." Language is a product of, and reflects, thinking. Sloppy usage reflects sloppy thinking, a kind of thinking incompatible with good scientific habits of mind.

By "usage" here I mean the technical term found in books on language that applies to choosing the correct word or form or tense rather than some approximation that the audience must parse for meaning. My students often chafe at my insistence on invoking the conventions of the

English language—it is just so "conventional," which connotes to them conservative, boring, and definitely un-hip. But language is, by definition, conventional—it is the set of definitions and rules by which we communicate, and communication is most definitely facilitated by conventions. A shiny purple vegetable is called an eggplant by convention even though it is not an "egg" (and doesn't even look much like one, especially if it is of the long, skinny, light-purple Japanese variety), and it is not a "plant" but the fruit of a plant; we have agreed, by convention, to call it an eggplant, and when I ask the grocer where I might find one, we achieve effective communication because we agree on the convention.

Although the plethora of poor usage examples I hear and read every day would require a second volume, I do think it worth recording here a few examples related to numbers and quantities, since these often arise when scientific thinking is involved. Perhaps the most common misuse that contributes to this particular pompous pedant's potful of pique are the ubiquitous supermarket signs that tell customers that an express checkout lane is reserved for those who have ten items or less. "Less" and "more" refer to "amounts" of continuous quantities (or quantities granular on such a small scale that the granularity is irrelevant): more or less milk, more or less sugar, more or less suffering in the world. "Greater than" or "fewer than" (and *please*, not "fewer then") refer to collections of discrete objects such as items in a grocery cart—one either has ten items or fewer than ten items (in which case you are allowed to use to the express checkout lane). "Amount" and "number" are similarly differentiated by referring to continuous and discrete quantities, respectively. It is not the amount of people in the world that is a problem but the number of people (as all the living ones, at least, are discrete quantities—and the ones that are in the process of decaying into the soil such that only some pieces are left are actually beneficial). Your tree could produce a large number of apples from which you might produce a large amount of applesauce (or, more cleverly, apple cider).

Another common usage error involves attaching modifying adjectives to words whose definitions do not allow modification. The most frequently abused example is the word "unique." The word derives from the Latin *unicus* and means singular, one of a kind, without equal. Something or some event is either unique or it is not. "Almost unique" means the opposite of unique since if it is only "almost" unique, it is, by definition, not unique (the correct choice here might be "highly unusual"). "Very unique" has twice as many words as required; there are no gradations of uniqueness—something is either unique or it is not.

There are other words related to mathematical concepts that, strictly speaking, should also not have modifiers: infinite, equal, singular, and perfect among others. A quantity is either infinite or it is not; nearly infinite means it is not infinite, just very large. Likewise, one quantity equals another or it does not, although I strongly suspect the battle against "nearly equal" has been lost. But the statement, for example, that there is almost an infinite amount of people in China still evokes my opprobrium as symptomatic of sloppy thinking. Granted, there *are* a very large number of people in China—nearly 1.4 billion to be precise. But no matter how large that number is, it certainly isn't infinite.

Some similar words have distinct definitions and are not interchangeable. "Farther," for example, refers to physical distance: the Earth is *farther* from the Sun in the summer; as for those who think we are closer in the summer, nothing could be *further* from the truth—"further" is used for metaphorical distance, including distance in time.

A single measurement of some phenomenon is a datum; multiple measurements form data, and data *are* important in science.

SCIENCE USAGE

Scientists and engineers are not, in general, renowned for their outstanding communication skills. (Do you know how you identify an extroverted engineer? He's the one looking at your shoes when he's talking to you.) But in addition to any generalized lack of language skill or personality anomalies, communication can be impeded by the lack of agreement on the conventional definition of words. I first became aware of this in the first class I taught at Columbia, a physics for poets class (most of them weren't actually poets, but they certainly weren't physicists either). When introducing light waves, I kept saying how they "propagated" through space. Met with uniformly blank looks, I tried another drawing on the board and a slightly different presentation but still used the word propagate, which to a physicist talking about a wave means "transmitted continuously through space" or some medium such as water. Finally, a student pointed out to me that her only acquaintance with the word *propagate* had to do with the production of baby rabbits. Since then, I have been more careful to define my terms when introducing a new concept.

Many words have multiple definitions. Of relevance here, however, is that a number of words have one definition that is commonly used in everyday speech but a particular, different meaning in science. In some cases, the common usage is technically wrong but has such wide currency that it is too late to recapture the convention. For example, saying a sack of flour weighs 1 kg is technically incorrect, since a kilogram is a unit of mass, not weight. The bag would have a mass of 1 kg even if it were floating weightless in a space station, because mass is a measure of resistance to a change of motion, and tossing the sack of flour to one's fellow astronaut would require the same force as it would on the ground. The conversion 1 kg = 2.2 lb is incorrect then because kilograms are a measure of mass and pounds are a measure of weight. The English unit for mass is the "slug"; 1 slug has a mass of 14.59 kg. And believe it or not, 12 slugs = 1 "blob" (= 175.126 kg—I'm not making this up; it makes a marvelous argument for adopting the metric system: 1,000 mg/g, 1,000 g/kg, 1,000 kg/ton is a little simpler). But now you see why even scientists usually let the incommensurability of pounds and kilograms slip by.

In everyday speech, speed and velocity are interchangeable, although to a physicist, speed is defined as how fast some distance is covered, whereas velocity includes the direction of the motion.

Secular trends are those that take place over time (they have nothing to do with a lack of church attendance), whereas secular changes that move in only one direction—e.g., rising sea levels over the past 100 years—are said to be monotonic (which doesn't mean they give rise to a single tone).

Good scientific models make predictions—that is, they forecast the outcome of an experiment or observation. In some cases, however, it is useful to test a model against some known data—can a climate model successfully postdict (after the fact) the amount of cooling induced by a volcanic eruption? Postdictions (which my spell-checker rejects) are useful for seeing if a new model is at least consistent with known facts but score many *fewer* points in a model's favor than predictions do. I emphasize the important distinction between a posteriori and a priori statistical reasoning in chapter 7.

Neologisms—newly minted words—are not uncommon in science because we are learning new things and inventing new concepts all the time. We call our impact on climate "anthropogenic" because, as I will demonstrate in chapter 10, it has its genesis in human (anthro) activity; likewise, as noted, geologists have adopted the term Anthropocene for the present—an age dominated by the presence of humans on Earth. There is even a Research Center for Scientific Terms and Neologisms in Athens that every few years publishes a list if you are intent on keeping up to date (although, since the last version was published in 2009, it, too, may have been a victim of the Greek debt crisis).

Just like numbers, words matter. They contain information and, when strung together correctly, the amount of information is amplified. When used sloppily or illogically, they simply contribute to the meaningless static of the Misinformation Age.

6
EXPECTING THE IMPROBABLE

"Attention passengers! This train is being held in the station due to police activity at 59th Street; someone has apparently fallen onto the tracks. We will be moving as soon as possible."

Thus began one of my most striking and memorable encounters with probability—an important object lesson in how abstract mathematical concepts can infiltrate everyday life. I had just been asked to write an article for a science magazine on an astronomical controversy concerning quasars. The standard model, described briefly in chapter 5, holds that quasars are very distant sources of enormous power driven by the accretion of material onto a massive black hole. This model invokes extreme conditions—a black hole of a billion solar masses or more, its ingestion of more than 100 billion trillion tons of matter each second—but the laws of physics used to describe these process are standard ones that have been tested in Earth-bound laboratories and through other astronomical observations.

Some time ago, however, there was an alternative hypothesis: that quasars are relatively nearby objects ejected from neighborhood galaxies. Their proximity, astronomically speaking, meant they had to produce far less power to appear as bright as they do. This made their energy requirements much less extreme but at a significant cost: to explain the large redshifts evident in their light, "new" (and unspecified) physics would need to be invoked. The principal advocate for this view was Dr. Halton Arp. In support of his unconventional viewpoint, he produced striking discoveries of quasars aligned with nearby galaxies, strongly suggesting a physical association between the two. Furthermore, Arp found improbable numerical relationships among adjacent objects. In one of his celebrated examples, two triplets of quasars were found lying in straight lines on the sky with separations in the ratio 1:3:4. His claim was that these alignments were too improbable to be viewed as merely random coincidences and so must represent some new and unknown physics.

I must confess to being quite impressed by this sort of coincidence and was pondering its potential significance for my article's conclusion as I walked to the subway station near my apartment to await the train to Columbia. At the first stop after boarding the train, I heard the announcement that opens this chapter. Later that day, as I was riding the Staten Island ferry across New York Harbor on my way to a weekend at my mother-in-law's house, I thought about the probability of the subway accident: what was the likelihood of my arriving within fifteen minutes of such an event?

In 2013, the New York City subway system recorded fifty-three deaths of people who fell, jumped,[1] or were pushed in front of trains from one of the system's 468 platforms.[2] If we make the simplifying assumption that all platforms and all times of day are equally risky, the probability of my arriving within fifteen minutes of such an event involves the number of events per year, the number of subway stations in New York, the number of visits I make each year to subway platforms, and the average time per day I am in the subway system. The answer is about 1 in 424 (see box 6.1). This means I would have to live in New York City for about 294 years to have a 50:50 chance of witnessing one such event. Yet I had just done so after less than a decade!

Approaching the Staten Island ferry terminus on my way back to Manhattan two days later, the train I was riding on stopped for an unusually long time at one of the stations. Someone had just fallen under the southbound train on the adjacent track, and heavy equipment was being called in to free the victim. The probability of my arriving within fifteen minutes of two accidents over three days of visits to the subway system in a given year is approximately 4.7 10^{-8} (or 1 in 1/4.7 10^{-8} = 21.3 million).

Now suppose you are on a jury and are listening to a prosecutor make the following argument: "The probability of the defendant (me, in this case) being present at two subway deaths in three days is less than 1 in 20 million . . . unless, of course, he is the perpetrator!" Would you vote to convict?

The purpose of this chapter is to equip you with the skills to make such a judgment. It may also help you decide whether to buy lottery tickets, provide you with a quantitative basis for skepticism about claims made in the media, and arm you with the tools and the confidence you need to debate your doctor; it even offers the opportunity to profit modestly at the expense of your friends who have failed to read this chapter.

DEFINING PROBABILITY

Probability and its sister discipline statistics have a terrible reputation that has endured for centuries. The epithet from which the title of the next chapter is derived—there are lies, damn lies, and statistics—coined by the nineteenth-century British prime minister Benjamin Disraeli, evinces a particularly robust notion: that probability and statistics are used primarily as tools for manipulation and deception. In addition, the two have a reputation as being abstruse, boring, or both. I would ask, however, that you suspend your belief in these cultural biases for the next two chapters so that I might attempt to win you over. For probability now shapes our understanding of the physical world, and statistics stands as the arbiter between our theories and the observations we use to test them. They are core habits of a scientific mind and provide a bulwark against skullduggery and exploitation. Indeed, they are essential tools for surviving the Misinformation Age.

Probability provides the basis for modeling phenomena that we think of as random. Every probability model has two components: the set of all possible outcomes and the number of outcomes of interest. By convention, probabilities run from zero to one, with zero denoting impossibility and one equaling certainty (the probability associated with the whole set of possible outcomes then, by definition, is equal to one). The probability (or "the odds" as one might say) of one outcome when there are only two equally likely possibilities is therefore 1 out of 2 or 1/2. A common instance of this situation is flipping a coin; the chance of getting heads is 1/2 or 50 percent because there is one outcome of interest (heads) out of two possible outcomes (heads and tails). As a warning that probability is not a completely intuitive concept, however, consider the case in which one desired outcome is only half as likely as the other. In this case, the probability is 1/3 (there are three possible outcomes, the one you want and the two—the twice as likely ones—you don't, so desired outcome = 1 and total outcomes = 3, yielding a probability of 1/3).

Probability calculations are particularly easy when the set of possible outcomes is finite and all the outcomes are equally likely; in that case, the probability assigned to the desired subset of the possible outcomes is simply the number of outcomes in that subset divided by the total number of possible outcomes. For example, the probability of pulling a king from a deck of fifty-two playing cards is calculated as follows.

The outcomes of interest are four: king of clubs, king of diamonds, king of hearts, and king of spades; all satisfy my desire for a king. The total number of possible outcomes in pulling one card from a deck of fifty-two different cards is fifty-two. The probability of getting a king then is 4 divided by 52, or 1/13 (1/13 = 0.077 or 7.7 percent).

Likewise, the probability of someone you meet having a birthday in the same month as yours just requires taking the number of outcomes of interest—one, your birth month—and dividing by the number of possible outcomes—all possible months, or twelve. So the probability of a matching birth month in $1/12$ = 8.3 percent (very similar to the odds of pulling a king).

In reality, this latter calculation is not perfectly accurate because there are not the same numbers of days in each month or the same number of births recorded on each day of the year. In fact, using data from the Centers for Disease Control and Prevention for 2012 and 2013,[3] I found there is a consistent trend of 5 percent more births per month in the second half of the year compared with the first six months. But even if births were distributed evenly over the year, the different month lengths would slightly modify the calculation. The chances of a February birthday would be $28/365$ = 0.0767 (in three out of every four years at least and $29/366$ = 0.0792 in leap years), whereas in March the odds would be $31/365$ = 0.0849. None of these is exactly $1/12$ = 0.0833, but they are all within a few percent of that value and provide a good quick estimate (remember from chapter 4 that estimates are often good enough).

SIMPLE PROBABILITY RULES

Three simple rules about combining probabilities provide a powerful set of tools to compute the likelihood of various outcomes in many situations. These rules are as follows.

Rule 1. The probability of the occurrence of an outcome from one of two mutually exclusive sets of outcomes is the sum of the probabilities of those two outcomes. In mathematical form, for mutually exclusive sets of possible outcomes A and B, $P(A$ or $B) = P(A) + P(B)$. "Or" is the key word here—if one outcome *or* the other are both acceptable, the probability of an acceptable outcome is the sum of the probabilities of the individual outcomes. This rule reflects an intuitive notion of how probability works. If we think of probability as "the proportion of time that an outcome will occur," the rule says that as long as A and B do not overlap, then the proportion of time that the two comprise in total is the sum of their respective proportions. For example, the probability of pulling a king or a queen from a deck of cards is $P(K \text{ or } Q) = P(K) + P(Q) = 4/52 + 4/52 = 2/13$, and the probability of a new friend having the same birth month as either you or your mother (assuming her birth month is different from yours) is $P(\text{yours or mother's}) = P(\text{yours}) + P(\text{mother's}) = 1/12 + 1/12 = 1/6$. This rule is consistent with our definition of probability, wherein the sum of all possible outcomes has a probability of one (100 percent). For example, the probability of getting a five when rolling a die is, by our definition, the number of outcomes of interest (one, a five) divided by the total number of possible outcomes (six), or 1/6; indeed, the probability of having any face of the die visible after a roll is 1/6. The probability of getting a 1 or a 2 or a 3 or a 4 or a 5 or a 6 is $1/6 + 1/6 + 1/6 + 1/6 + 1/6 + 1/6 = 6/6 = 1$.

Rule 2. The total probability of two independent events *both* happening is the product of the probabilities of the events. Mathematically speaking, for two subsets of possible outcomes, A and B, $P(A \text{ and } B) = P(A) \times P(B)$. Here the key word is "and"; we need both A **and** B to happen if we are to be satisfied and are looking for the probability of both events occurring. The other key word is "independent": how do we know if two outcomes are truly independent? One answer is that the product formula above is the *definition* of independence, and for mathematicians interested only in the mathematical consequences of various definitions, this suffices. But we also have an intuitive sense of what we mean by independence—the occurrence of one event will have no influence on how likely another event is to happen. For example, the probability of the first two people you meet today both sharing your birth month is $P(\text{both sharing your month}) = P(\text{first shares}) \times P(\text{second shares}) = 1/12 \times 1/12 = 0.0069$, less than one chance in a hundred (assuming, of course, that your first two acquaintances are not twins). If this actually happened to you, you might well find it a remarkable coincidence, just as Dr. Arp did when he found all those aligned quasars, and you might even attach some cosmic significance to the event.

(If that would be your inclination, be sure to read the section later in this chapter entitled "Rare Things Happen All the Time.") Returning to our deck of cards, does this rule mean that the probability of pulling a king *and* a queen from a deck of cards $1/13 \times 1/13 = 0.0059$? Not quite.

It is essential in calculating probabilities to make certain that the question you are asking is clearly stated. Are we interested in the probability of first obtaining a king then obtaining a queen on two successive draws? Or in the probability of holding a king and a queen after two draws? It is important to keep track of both the set of possible outcomes and the individual probabilities. For example, starting with a deck of fifty-two cards containing four kings and four queens, you draw one card at random and then another from those remaining in the deck (we are assuming all possible sequences of draws are equally likely). Now suppose that your goal was to draw first a king and then a queen. As shown previously, the first draw has four possible outcomes of interest (one of the four kings) and fifty-two total possible outcomes. For the second draw, however, although there are still four outcomes of interest (the four queens), there are only fifty-one cards from which to choose. Thus, the probability of pulling a king and then a queen is given by $4/52 \times 4/51 = 16/2652 = 0.00603$ or 0.603 percent.

Not quite $1/13 \times 1/13 = 0.00592$. Why doesn't the product rule for independent events hold here? Why can't we simply multiply the probability of getting a king on your first draw, $1/13$, by the probability of getting a queen on your second draw, $1/13$? The reason is that the events are not completely independent: if you get a king on your first draw, there is now a $4/51$ rather than $4/52$ probability of getting a queen on the second draw because the act of drawing the king has reduced the total number of possible outcomes (the number of cards in the deck) to 51.

If instead of insisting on a king first and a queen second your goal were simply to end up holding a king and a queen after two draws, the probability can be calculated by breaking the event up into two mutually exclusive events. One possibility is that you draw a king first and a queen second, whereas the other acceptable outcome is to draw a queen first and then a king. The probabilities of these two events are the same as shown previously. Since either the first or second are acceptable—in either case we'll end up holding one king and one queen—we simply apply rule 1 and add the exclusive events: 16/2652 + 16/2652 = 32/2652 = 0.012 or 1.2%. We can extend both of these rules to more than two possibilities. The extended version of rule 1 is that for multiple mutually exclusive subsets, $P(A_1$ or A_2 or A_3 or . . .) = $P(A_1) + P(A_2) + P(A_3)$. . . . The extended version of rule 2 is that for multiple mutually independent events, $P(A_1$ and A_2 and A_3 . . .) = $P(A_1) \times P(A_2) \times P(A_3)$. . . .

Rule 3. A third rule is implied by our definition of probability: the probability of something not happening is equal to one minus the probability of it happening, $P(\text{not } A) = 1 - P(A)$. This rule follows from the addition rule for mutually exclusive events (rule 1) and the convention that the probability of the set of all possible outcomes is one, since event A has no overlap (is mutually exclusive with) the event "not A." In other words, the third rule simply states that the probability that one outcome *will* occur is equal to one minus the probability that the outcome *will not* occur.

PROBABILITY PAYS

What is the practical application of all this? The mathematical theory of probability had its origins in the 1650s in discussions about games of chance, so how about applying the knowledge to make money off friends and acquaintances? Or, in my case, my students. For the past twenty-five years, I have attempted to reinforce my lessons in probability by lightening the wallets of my students using the following demonstration.

Let's say there are sixty students in the class. There are 366 possible birthdays for those sixty students (although February 29 occurs only

25 percent as often as the other 365). I know none of their birthdays. I offer odds of 20:1 that at least two will have the same birthday and emphasize the point by laying $20 bills all across the lecture table at the front of the room and inviting people to cover the bets with $1 each.

It sounds like an almost irresistible deal—only sixty total birthdays and 366 days from which to choose. It seems as though I should be demanding 1:6 odds (i.e., 60/366) in my favor rather than giving 20:1 odds to the class. But probabilities are often not what they "seem" to be. Let's follow the previous definition and rules and calculate the odds of two people having the same birthday.

In this problem, as in many others, it is often most convenient to frame the calculation backwards, i.e., to ask what the probability of people having different birthdays is. As noted previously, the probability of two people having the same birthday is one minus the probability of each having a different birthday, or P(two identical birthdays) = 1 − P(different birthdays).

Let's start with the first person. The odds of her having her own birthday is one. For the second person, the odds of having a different birthday than person 1 is, ignoring leap years to keep it simple, 364/365 (there are 364 outcomes of interest—a different birthday—out of 365 possible outcomes). With two dates now taken, the odds of a third person having yet a different birthday is 363/365, and so forth. If we want the birthdate of person 2 to be different from the birthdates of persons 1 and 3, we—according to rule 2—we multiply the probabilities. Thus, the probability of two people in any random group having the same birthday is P(two identical birthdays) = 1 − [1 × 364/365 × 363/365 × 362/365 . . . × (365 − n − 1)/365], where n is the total number of people in the group. If you carry this out, you will find that for n = 23, the odds are roughly 50:50 that two will have the same birthday. Yes, only twenty-three! So while you may find it amazing when you discover that two people at a party have the same birthday, if there are more than twenty-three people at the party, probability *predicts* that this should happen more than half the time. In my class of sixty, the odds are P(two identical birthdays) = 1 − 0.0059 = 99.4 percent, or nearly 200:1 in my favor—which explains why I have never lost in twenty-five years of fleecing my students.

The message here, in addition to an obvious money-making scheme to use at your next party, is that your common sense or gut feeling about the likelihood of some "coincidences" can be very misleading. Beware.

BOX 6.1 THE ODDS OF SUBWAY ACCIDENTS

With our three rules in place, we are now in a position to calculate the subway accident probabilities that I cited at the beginning of this chapter. If we assume two visits to the subway per day, there are 2 visits per day × 365 days per year × 15 station minutes per visit = 1.095×10^4 station minutes per year during which I could be delayed by a subway accident. And there are a total of 365 days × 24 hours per day × 60 minutes per hour × 468 stations = 2.460×10^8 station minutes in which these events could occur. If we make the simplifying assumption that an accident is equally likely to occur at any given time and station, then the probability that a particular accident causes me a delay is $1.095 \times 10^4 / 2.460 \times 10^8 = 4.45 \times 10^{-5}$ or 1/22,464.

There are fifty-three such accidents in a year, and we will assume (reasonably) that they occur independently. The probability that at least one of the fifty-three accidents causes me a delay is one minus the probability that none of the accidents cause me a delay, which equals one minus the probability that any of the fifty-three accidents cause me a delay multiplied by itself fifty-three times. That is, $1 - (1 - 1/22{,}464)^{53} = 0.00236$, or 1 chance in 424 in a year. If I want to know how many years I must ride the subway in order to have a 50:50 chance of being delayed by an accident, we note that the probability of not being delayed in n years is $(1 - 0.0024)^n$, and solving $(1 - 0.0024)^n = 0.5$ gives $n = 294$. If you wish to test your grasp of what we've done so far, review the previous calculations and note which rule (first, second, extended first or second, or third) underlies each step.

CONDITIONAL PROBABILITY

So far we have discussed events that are strictly independent of each other—cards pulled from a deck, birthdays for unrelated people, and accidents on two distant subway platforms can in no way affect each other, so their probabilities of occurrence are independent. There are many real-world situations, however, where this is not the case. One class of situations that is important to almost everyone at some point in their lives is medical testing. Understanding these situations requires us to introduce the notion of *conditional probability*.

When two events are not independent, the occurrence of one has implications for the other. For example, the probability that a lab test for a given disease will come back positive is much higher if the patient has the disease than if he or she does not. However, virtually all medical diagnostic tests sometimes return "false positives"—that is, the test results indicate the patient has the disease when in fact he or she does not. To correctly interpret the test results, one needs to know three things: (1) the test's efficacy or the probability that the test yields a positive result when the patient actually has the disease; (2) the false-positive rate or the probability that the test yields a positive result when the patient does *not* have the disease; and (3) the prevalence of the disease in the population as a whole or the probability that any given patient has the disease.

What we really want to know, of course, is which patients have the disease so we can initiate treatment and which do not so we can send them on their way. Given the imperfections of most diagnostic tests, however, we rarely have certainty (probability = 1). What we must calculate is the conditional probability of a patient with a positive test result actually having the disease by computing a ratio of probabilities: the probability of being sick and having a positive lab test divided by the probability of the positive test result. Expressed more generally, the conditional probability of an event A occurring, given that B occurs, is computed as $P(A$ and $B)/P(B)$.

We can turn the example around. Imagine that you are a healthcare worker presented with a patient with a positive lab test. Suppose that the lab test is 95 percent reliable: i.e., the probability of a positive outcome given the presence of the disease is not 100 percent (as it almost never is) but only 95 percent. And suppose that the probability of a positive result among disease-free subjects—the false positive rate—is 5 percent. Finally, suppose that the prevalence of the disease in the population is quite low— only 1 percent of individuals actually have the disease. As a medical professional, you are concerned about both the cost and the side effects of treating the patient—especially if the test result is a false positive. It is important to know how likely it is that your patient really has the disease given the positive lab test results; in other words, you want to know the conditional probability that the patient is sick.

According to our definition of conditional probability, we need to compute (1) the probability of having the disease and testing positive, (2) the probability of testing positive, and (3) the ratio of these two quantities.

We are given that 1 percent of the population has the disease and that the test is 95 percent effective at yielding a positive result when administered to truly sick patients. These are independent probabilities because 1 percent of people would have the disease whether the test existed or not, and the test is right 95 percent of the time whether the patients exist or not. So, as before, to calculate the probability of two independent events, we just multiply the probabilities of the two events: in the first calculation, P(sick and positive) = 1 percent sick × 95 percent positive when sick = $0.01 \times 0.95 = 0.0095$. Now, the probability of testing positive comes from two mutually exclusive events: either being sick and testing positive or being well and testing positive. What we want is the total probability of a positive result P(sick and positive) + P(well and positive). We just calculated the first of these as 0.0095. The second calculation comes from knowing that 99 percent of the population is well and that the test gives a false positive 5 percent of the time, so P(well and positive) = 99 percent × 5 percent =

0.99 × 0.05 = 0.0495. Thus, P(positive) = P(sick and positive) + P(well and positive) = 0.0095 + 0.0495 = 0.059.

With these two calculations we have our ingredients for the conditional probability that the patient with the positive test result is actually sick—it is just the ratio of these two numbers: P(sick and positive)/P(positive) = 0.0095/0.059 = 0.16 or 16 percent, meaning that only 16 percent of the patients testing positive are sick at all. Treating them all would be a waste of resources and produce completely unnecessary side effects in five out of every six patients. Perhaps you ought to order another independent lab test for confirmation.

Note that the numbers that characterize the test in this example are not unreasonable for real clinical practice. Even if we make the test much better—say it is 98 percent accurate when the patient is really sick and has only a 1 percent false positive rate—the fraction of patients with positive tests who are not sick still exceeds 50 percent.

A key factor here is the prevalence of the disease in the population; the rarer the disease, the worse the false-positive problem becomes. If the disease is common—say 50 percent of the population has it—then even a somewhat flawed test such as the one in the previous example is right 95 percent of the time, although, of course, you'd be right 50 percent of the time by simply guessing.

RARE THINGS HAPPEN ALL THE TIME

I propose now to convince you that this section title is not oxymoronic (or just plain moronic). Along the way, I will try to induce some healthy skepticism informed by your growing knowledge of probability.

The human mind has evolved over the past few million years to look for patterns. There are many instances in which this highly developed ability is important for the survival of an individual or of the species. The ability of the eye-brain combination to quickly recognize tiger stripes is undoubtedly an advantage to an individual who wishes to pass on his genetic code to future generations (see chapter 5). The recognition of recurrent patterns in the length of the day, the temperature, and the amount of rainfall clearly enhances the ability of a society to hunt more efficiently or even to plant crops for food. But in an age of caged tigers and Krispy Kreme doughnuts, some of our primitive cognitive skills are no longer essential for survival, and their application in a modern technological society can lead us astray. Indeed, our predilection for seeking patterns can make truly random events seem highly ordered and in need of an "explanation."

In fact, humans have a remarkable penchant for "explaining" random events, as well as for accepting nonrandom events at face value. Both tendencies can get you into trouble. Take, for example, a simple (and oft-used) stock swindle.

Suppose tomorrow morning you receive an e-mail from me with a stock market prediction: over the next two weeks ticker symbol HAH is going to move up. You watch the stock out of curiosity and, as it turns out, I'm right. In a fortnight you get another tip from me: keep watching that stock—it's going up again over the next two weeks. It does. Two Fridays hence my e-mail tip sheet is there again to let you know that, according to my analysis, HAH will experience a significant selloff over the next two weeks. And it does!

This pattern continues for another two months; every two weeks you get an e-mail that predicts the direction of the stock's movement, and seven times in a row I am right. The eighth message is slightly different, however. It asks you for a $300 annual fee to keep receiving these tips. That sounds cheap given that, by investing $1,000 in HAH and following the tips I provided, you could have cleared $800 over three months; you sign up right away. But if I am so good at this, a skeptic might well ask, why am I writing this book instead of sipping martinis (shaken, not stirred) in the Seychelles?

Your mistake, you see, is in believing there is information behind my predictions that allow me to know how the stock will move. After all, the probability of getting it right seven times in a row by chance is low. Since I am just making a binary prediction—up or down—it is equivalent to flipping a coin and getting seven heads in a row. And you know how to calculate that probability: it's $\frac{1}{2} \times \frac{1}{2} \times \frac{1}{2} \times \frac{1}{2} \times \frac{1}{2} \times \frac{1}{2} \times \frac{1}{2} = 1/128$, or less than 1 percent. In fact, however, it wasn't hard at all to get it right seven times in a row, and it required precisely zero real information.

The first week, I sent the e-mail to 2,000 people. In half of the letters I said HAH would go up; in the other half I predicted it would decline. It went up. I discarded the addresses of the 1,000 people to whom I had predicted a decline; for the other 1,000 people to whom I sent a new e-mail, 500 predicted a continued rise and 500 claimed a selloff was imminent. For the third tip sheet, I used just the 500 addresses of people for whom I was two for two and sent 250 positive and 250 negative prognostications. After seven fortnightly predictions, there were about thirty-two people who were convinced of my market clairvoyance and 1,968 who had stopped hearing from me.

In the days of postage stamps these mailings would have cost $1,739 (at $0.45 for a first-class letter—and this is clearly a first-class operation). But if just six of the thirty-two people who witnessed seven straight correct predictions bought my newsletter, I made a profit. If half did, I pocketed $3,000. Today, of course, postage is not required, and a list of 2,000 could easily become 2,000,000 with a little extra effort. This is why your e-mail program has a spam filter. But it's not perfect, so having your own spam filter continuously running in your prefrontal cortex is an excellent idea. HAH!

To be more precise about the title of this section, I really should have parsed it as follows: (1) improbable things happen all the time, and (2) rare things happen. As an example of the first, take the New York State lottery. The odds against winning when you buy a ticket vary depending on the number of tickets sold but are usually at least five million to one. What a remarkably improbable event winning is—one chance in five million. It's almost as improbable as being present at two subway deaths in three days without causing either of them. And yet, of course, it happens every week. Someone (or often more than one someone) wins. How can we call a daily event improbable?

The probability of something happening must, as noted previously, be carefully defined. The odds of *you* buying a ticket and winning may be five million to one, so if it happens, you are "amazed." Note, however, that you have calculated the odds of this amazing event *after* it has happened. What you have failed to note is that the odds are also five million to one for everyone else buying tickets, and if ten million tickets are sold, it is hardly surprising that someone wins. Improbable things happen all the time.

An important issue here is the difference between a priori and a posteriori calculations. These Latin phrases simply refer to calculations done before and after the fact, respectively. Failure to recognize a vast distinction between the two is a major source of confusion and misinformation in the world. And it is key to understanding my second point: rare things happen. Consider the following rare and apparently wildly improbable event.

Your name is Brianna. You were born on February 23, and you have a brother named Jonathan. Your parents' names are Mary and Dick. You were born in Minneapolis, went to college there, and have managed to land a (paying!) position with a major publisher in Chicago.

Monday is the first day at your new job. Understandably, you are a little nervous and anxious, among other things, about getting to know your coworkers. Your new boss squires you around the office to introduce you; she smiles as she introduces you to the person in the adjacent cubicle—her name is also Brianna. She's only the second person you've met with your name, and you think it is a little odd the boss decided to seat you next to one another, but this other Brianna seems nice enough and you agree to go to lunch with her the next day.

At the local sushi joint you have the usual ice-breaker conversation but are amazed at what you discover about each other: her birthday is also February 23 and yes, you guessed it, she has a brother named Jonathan too.

Clearly, cosmic forces are at work (after all, you are both the same astrological sign). But let's calculate the odds of this happening by chance.

Conducting a Google search for "frequency of girls' names," I immediately found some useful data. A recent sample of 85,000 given names found that of the 41,465 girls born in a single year, 232 were named Brianna (1 in 179), while of the 43,740 boys born in the same year, 341 were named Jonathan (1 in 128). So the probability of this event is calculated as follows: P(you being named Brianna) = 1 (that's your name!); P(another random female you encounter named Brianna) = 1/179; P(your brother being named Jonathan) = 1 (that's his name); P(your officemate's brother being named Jonathan) = 1/128; P(your birthday being February 23) = 1; and P(your roommate's birthday being February 23) = 1/365.

Thus, the probability of all these independent things being true at the same time is given by their product $1 \times 1/179 \times 1 \times 1/128 \times 1 \times 1/365 = 1.2 \times 10^{-7}$—nearly a one in ten million chance! Even rarer than winning the lottery. It must be in the stars. Or must it?

There are now 320 million people in the United States, of whom 246 million are over sixteen years old, and 155.4 million of those are employed. The Bureau of Labor Statistics reports that the average worker today holds a job for 4.4 years (although millennials seem to change jobs more frequently).[4] That means, each week, roughly 680,000 people find themselves in the situation of starting a new job and finding out their coworkers' names; a large fraction is likely to also discover at least one colleague's birthday and brother's name. If we want to know how likely it is that the exact outcome I have postulated occurred, we simply multiply the probability of it happening by the number of times we perform the experiment: $1.2 \times 10^{-7} \times 6.80 \times 10^5 = 8 \times 10^{-2}$, or one chance in twelve—*every week!*

However, you would no doubt find the coincidence of first names, birthdays, and brother's names equally astounding if it were two Emilys, both born on December 7 and both with brothers named Jacob. In fact, in the sample I used, Emily and Jacob were the two most common names, and the odds of that happening are therefore higher—about one in two. To find the probability of this triple coincidence occurring for any set of names, all I need to do, according to our rules, is to add the probabilities of each set of names. In the end, many such coincidences will happen each week. In a strictly random universe bereft of astrological influences, such a remarkable event will certainly occur.

Yes, it is a rare event, but no, it does not require an explanation—we should *expect* such coincidences to happen. If it happens to *you* . . . yes, well, it has to happen to somebody.

Confronted with a new colleague of the same name and birthdate with matching sibling names, you might well do the first calculation above and tout it as a one in ten million chance. But suppose it were your mothers rather than your brothers with the same name? Or your fathers? Or uncles? Or dogs? There is a very large number of "coincidences" that *could* have occurred. In this case, you have first defined the exact set of coincidences and then have done the calculation—a clear case of a posteriori statistics. If, on the other hand, you defined in advance all of the coincidences you would describe as remarkable and then added the probabilities together ("I would find this or this or this . . . strange"), the chance of *something* strange happening would be appropriately judged as much more likely.

In science, we place far greater value on a priori calculations than on a posteriori ones. As noted at the beginning of this chapter, Halton Arp found "remarkable" coincidences between quasar separations and positions, and the longer he looked for such events, the more spectacular the coincidences became. But we should *expect* this even if there is no physical connection at all between the quasars and nearby galaxies. If one flips a coin a hundred times, it is very unlikely that ten consecutive heads will come up. The odds against this are, according to our rules for combining independent events, $\frac{1}{2} \times \frac{1}{2} \times \frac{1}{2} \ldots = (\frac{1}{2})^{10} = 1$ in 1,024, so in ninety-one tries[5] there is less than a 10 percent chance of it happening. In a million flips, however, this "unlikely" event should be expected to occur. Eventually, the ardent coin flipper will cease to react with surprise to ten heads in a row but might still find "remarkable" a run of twenty. (What are the odds against that? Only about one in a million.)

The bottom line is this: it is important to define the rules of the game *before* beginning to play. Arp only reported his quasar–galaxy coincidences after he found them. Thus, although each new report was more stunning than the last, his hypothesis of nearby quasars soon faded from the scientific scene; he never produced an a priori prediction, and his a posteriori statistics no longer impressed anyone.

A MATHEMATICAL MODEL OF LUCK

What is luck? Is it determined genetically? By the stars and planets? By zip code?

In my view, some of these explanations are more likely than others. But consider the following scenario based on a story in John Allen Paulos's charming book *Innumeracy*.[6]

Two people who lunch together every day decide that, instead of bothering to figure out how to divide each bill, they will flip a coin and keep a running total of heads and tails. If, on any given day, there are more heads, Brianna will pay, whereas if tails are in the lead, Jonathan will pay. If heads = tails, they'll simply split the bill in half. Notice this system is subtly different from one in which each day's flip determines who pays. But it still sounds like a fair enough system, yes?

Well, after three years, one of the two is going to think of herself as a real winner, and the other is going to demand an end to the practice. How do I know this?

After 1,000 flips, it is *much* more likely that one person pays more than 90 percent of the time than it is that they are splitting as closely as 45:55, and it is even more likely that one pays 96 percent of the time than they are as close to even as 48:52.

But how can this be? For a truly random process such as flipping a fair coin, you would expect the ratio of heads to tails to get closer and closer to the average, or a "mean" value of 50:50. One flip and it will be 100 percent heads and 0 percent tails, and after ten flips it may be 60 percent heads and 40 percent tails. But after 100 flips, you would not expect sixty heads and forty tails—probably more like fifty-four and forty-six, respectively. And these expectations are correct. They even have a name: "regression to the mean," sometimes referred to colloquially as "the law of averages." But suppose you flip six heads in a row. What is the probability of getting a head on the next flip?

If you have completely internalized and accepted the rest of this chapter, you will respond 50:50. There is *always* a one in two chance of getting heads (one outcome of interest over two possible outcomes). Intellectually you accept this. But do you *really* believe it? If you flipped eleven heads in a row, don't you feel strongly that the next flip is likely to be a tail?

This "feeling" also has a name: the gambler's fallacy. If you have lost several hands of blackjack in a row, you just know your luck has to change, so you keep playing. Casinos rely heavily on this feeling, which is largely responsible for the $38.8 billion in casino revenues in 2014.[7]

At first, you might appeal to regression to the mean: If I have eleven heads in a row, and I know if I keep going I have to get closer and closer to 50:50, then clearly it must be the case that a tail is more likely. Wrong! It is wrong to think of regression to the mean as a rubber band. This is why the lunch partners are sure to split in anger after awhile. Say the score is 519:481 for Brianna and Jonathan after a thousand flips. That's pretty close to 50:50—within 2 percent. But the odds of a head on the next flip are still 50:50—Brianna's advantage is just as likely to grow as it is to shrink.

It *is* true that in the next 1,000 flips, the ratio of heads to tails is likely to grow ever closer to 50:50. But the difference between heads and tails tends to grow with time, and the lead changes become less and less frequent. Brianna begins to feel like a lucky person—she believes there is such a thing as a free lunch since she gets one every day. Jonathan resigns himself to being a perennial loser, and the psychological effects can be self-reinforcing.

In his seminal 1948 article "The Self-Fulfilling Prophecy,"[8] the sociologist Robert Merton quotes a theorem enunciated by his colleague W. I. Thomas: "If men define situations as real, they are real in their consequences." If Jonathan defines his repeating, daily lunch debt as arising from some defect of character or simply that he was born "unlucky," real consequences can ensue. He may be more reticent about taking risks or interpret neutral events as negative—he may, in short, engage in a self-fulfilling prophecy of failure.

The antidote to this downward spiral is found in this chapter. Brianna's "luck" and Jonathan's lack thereof are simply consequences of purely random events described by the rules of probability. Like streaky shooting in basketball[9] and innumerable other examples that pass for "common knowledge" in our society, Brianna's "luck" is nothing other than the operation of completely random chance and has no predictive value whatsoever for any other situation in which she finds herself. Jonathan's "unluckiness" is equally devoid of meaning. It is all the product of totally random events that obey the laws of probability. Any "explanations" are thus completely bogus; inferences from such explanations can be dangerous and only contribute to the clutter of the Misinformation Age.

CONCLUSION

There is nothing inherently wrong with seeing patterns and looking for explanations in the world around you. As noted earlier, it is a natural consequence of evolution that our brains engage in such activities. But doing this compulsively—like my spouse's dog sniffing every parking meter, fire hydrant, and lamppost to see if another dog has passed recently—can be unproductive. Many "patterns" are illusory; finding explanations for them can lead both individuals and societies to adopt irrational beliefs.

Returning to the subway: I didn't push anyone onto the tracks. It's just that I noticed one of the many, many events (such as seeing two professional athletes on my way into the station, hearing two automobile accidents through the grate above the subway platform, buying two winning lottery tickets at the kiosk by the turnstile, etc.) that, had they occurred, would have seemed remarkable. But in all of these cases, my wonder at the event would be tempered by my knowledge of probability and the recognition of my brain's embedded algorithms that search for meaning even when no meaning exists.

Searching hard for explanations of events in the world around us is one of the first steps in scientific inquiry. A posteriori searching can suggest possible explanations, but these explanations must then lead to a priori calculations that, in turn, are to be tested against fresh data before any new explanation can be adopted as a scientific model. And we must always be cognizant of the fact that the event we are attempting to explain may simply be a manifestation of the random workings of the universe and requires no explanation at all beyond the simple laws of probability.

7
LIES, DAMNED LIES, AND STATISTICS

"Some notable murder cases have been tied to the lunar phases. Of the eight murders committed by New York's infamous 'Son of Sam,' David Berkowitz, five were during a full moon."[1]

The Son of Sam case introduced the term serial killer to our language. His capture on August 8, 1977, led, among other things, to my move to New York two weeks later (not really—I was coming anyway). The matter of interest here, however, is the fact that this news story from the BBC provides powerful supporting evidence for a common piece of folk wisdom: the phases of the Moon affect people's behavior.

Some years ago I was having dinner with a college friend in Sydney, Australia, where she now lives. Her twenty-something-year-old daughter was along. At one point, the conversation turned to behavior and the full moon, which, her daughter said, "everyone knows it makes people act crazy."

"No," I couldn't resist interjecting. "There is no evidence for that whatsoever. There have been dozens of studies in which carefully collected data were subjected to rigorous statistical analyses, and none has shown any connection."

"Well, you can prove anything you want with statistics," she said dismissively.

I must confess that I find her second statement more disturbing than the belief in disproven folk wisdom expressed in her first. One *cannot* prove "anything you want" with statistics if they are applied to reliable data in a logical and mathematically correct fashion. Indeed, statistical reasoning is essential for examining whether data provide significant and meaningful evidence for or against a scientific model. In the hands of a scientifically minded person, statistical methods are, more often than not, used to *disprove* what "you want"—support for your favorite novel idea.

A simple illustration of statistical reasoning can be applied to test the claim of this news story, a claim repeated hundreds of times during the Son of Sam's reign and countless times since. The dates and times of the murders (or shootings—only six, not eight as the news story had reported, of the victims died) were in fact as follows:

July 26, 1976 – 1:00 A.M.
October 23, 1976 – 3:00 A.M.
November 26, 1976 – 11:00 P.M.
January 30, 1977 – 00:10 A.M.
March 8, 1977 – 10:00 P.M.
April 17, 1977 – 3:00 A.M.
June 26, 1977 – 3:00 A.M.
July 31, 1977 – 1:00 A.M.

Now, the orbit of the Moon around the Earth has been *very* accurately determined—in fact, by bouncing laser light off little mirrors left on the surface by the Apollo astronauts and timing its return, we measure the distance to the Moon every day to a precision of roughly one millimeter. Since we also understand gravity—the force that keeps the Moon in its orbit—it is straightforward to calculate the exact position of the Moon, and thus its phase, at any point in the past or the future for tens of thousands of years. A nice calculator is readily available online;[2] you can use it to find the following phases for the dates given above: 28.46 days, 29.44 days, 5.83 days, 9.58 days, 18.26 days, 28.32 days, 9.06 days, and 15.41 days.

The accounting system labels the new moon—when it is perfectly aligned between the Sun and the Earth and is thus invisible to us—as 0.00 days (box 7.1). During the year in question, the Moon had an orbital period around the Earth (and thus an interval from one new moon to the next) of 29.52 days (this varies slightly over time as the Moon is tugged about by the Sun and the other planets). The full moon occurs halfway through the orbit at 14.76 days.

Using the data above, it is easy to calculate the number of days between the full moon and each attack by just taking the absolute value[3] of the difference between the given phase and 14.76 days; i.e., the first murder

---

BOX 7.1 PHASES OF THE MOON

Contrary to another bit of common—and incorrect—folk wisdom, the Moon does not have phases because the Earth is casting shadows on it. As it orbits the Earth over a month, it also rotates once on its axis, so that the same side is always facing the Earth. However, again skewering common misinformation, there is no "dark side" of the Moon (Pink Floyd notwithstanding). At the new Moon—designated "1" in figure 7.1—the Moon is between the Earth and the Sun and, as always, the side facing the Sun is lit and the other side is dark—this is the "new" Moon when we see no Moon in the sky. Seven days and a few hours later, the Moon has moved to position 3; in this configuration, we see a half Moon (one half of the lighted side and one half of the dark side). At position 5, halfway through its orbit at 14.76 days, the Moon is full. Also marked on figure 7.1 are the times of the Son of Sam shootings.

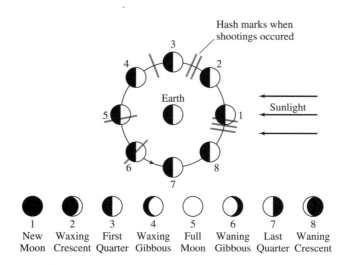

FIGURE 7.1 LUNAR PHASES

The phases of the Moon marked with the points at which the Son of Sam shootings occurred.

occurred $|28.46 - 14.76| = 13.7$ days from the full moon, whereas the third occurred $|5.83 - 14.76| = 8.93$ days from the full moon. Summing all these absolute differences and dividing by eight (the number of attacks) gives the average time between the closest full moon and the attacks: 8.24 days.

Now, if the attacks had all occurred near the full moon as claimed, the absolute values would tend to be small, and the average would be near zero. However, if they were distributed independently of the timing of the full moon (the anticipated result if there were absolutely no connection), we would expect the times to be evenly distributed between 0 and 14.76 days and thus, on average, to be $14.76/2 = 7.38$ days (see box 7.2). It is obvious that the true value is much closer to the value expected for no relationship than it is to the smaller values asserted by the "lunatic" theory (and the claims of countless media

---

BOX 7.2 DEFINING AVERAGE

The average (or mean) of a set of values is equal to the sum of the values divided by the number of values there are. The average of the real numbers x, y, and z is $(x + y + z)/3$.[4] The average of 0 and 14.76 = $(0 + 14.76)/2 = 14.76/2 = 7.38$.

If you take a uniform distribution of values (that is, equally spaced values) between 0 and 14.76, by definition half will be above the mean (7.38) and half below it. In other words, the mean of all the values between 0 and 14.76 will be the same as the mean of 0 and 14.76 alone.

Thus, if dates are randomly distributed between the new moon date (0 days) and the full moon date (14.76 days), we expect the random date to fall, on average, midway between these two dates, or at 7.38 days.

---

reports). Using statistical reasoning to adjudicate between competing theories is a hallmark of good science and a vital tool for surviving the Misinformation Age.

HOW *NOT* TO LIE WITH STATISTICS

Experiments have been a key component of science since the Renaissance when Simon Stevin dropped two weights from a tower in Delft and found that, contrary to two thousand years of Aristotelian dogma, they fell at the same rate independent of their masses.[5] Experiments produce *measurements* that are then compared to the predictions of models. An experiment has the power to distinguish between competing models only inasmuch as the competing models predict different measurements.

To compare models to measurements, the models should make specific predictions, and the experiments should yield either numerical results or the classification of outcomes into clearly defined categories. This section explores some of the rules scientists adopt when dealing with measurements and the numbers that describe them.

ACCURACY AND PRECISION

We begin by drawing a distinction between two words that are often used interchangeably in everyday speech but that in science mean two quite different things: accuracy and precision. On the archery range, for example, you might say someone shot with great accuracy or with great precision and mean more or less the same thing. However, as you probably learned as a child (if, like me, you grew up in this backward country that still resists the otherwise universal metric system), the average human body temperature is 98.6°F. To a scientist, this number is quite precise: quoting it to three figures implies we know it to within 0.1 degree out of nearly 100 degrees or one part in 1,000 (i.e., it is not 98.5 or 98.7 but 98.6). However, the number is *not* accurate.

First, 98.6 is a direct conversion of the Celsius scale value of 37°C; note that this is a whole number—the value in Celsius is only quoted to two digits since it is known to be just a rough average, whereas the Fahrenheit value is quoted to three (see box 7.3 for a discussion on significant figures). Second, the average temperature of a healthy person taken with an under-the-tongue thermometer is actually about 0.4 degrees lower, or 98.2°F. Furthermore, the range of "normal" temperatures among different individuals is at least 2°F; in addition, it is normal for one's temperature to vary by about 1 degree over the course of the day. Quoting the average human body temperature as 98.6°F, then, conveys misinformation about the precision with which this quantity is known.

BOX 7.3 DEFINING SIGNIFICANT FIGURES

The rules for significant figures are important because they convey directly to the scientific reader the precision of the measurement in question. These rules are as follows:[6]

1. *All nonzero numbers are significant.* The number 33.2 has three significant figures because all digits present are nonzero.

2. *Zeros between two non-zero digits are significant.* The number 2,051 has four significant figures. The zero is between a two and a five.

3. *Leading zeros are not significant.* Leading zeros are nothing more than place holders. The numbers 0.54 and 0.0032 both have two significant figures. All of the zeros are leading and don't count.

4. *Trailing zeros to the right of the decimal are significant.* There are four significant figures in 92.00. This is important: 92.00 is different from 92. A scientist who measures 92.00 milliliters is reporting her value to the nearest 1/100th milliliter and is implicitly claiming to know the value to that level of precision; meanwhile her colleague who measured 92 milliliters only knows his value to the nearest one milliliter. It's important to understand that zero does not mean nothing. In this case it denotes actual information just like any other number. You cannot tag on zeros that aren't justified by the quality of the measurement—that conveys misinformation.

5. *Trailing zeros in a whole number with the decimal shown are significant.* Placing a decimal at the end of a whole number is usually not done. By convention, however, this decimal indicates a significant zero. For example, the decimal point in "540." indicates that the trailing zero is significant; there are thus three significant figures in this value.

6. *Trailing zeros in a whole number with no decimal shown are not significant.* Writing just "540" indicates that the zero is not significant, and there are only two significant figures in this value.

7. *For a number in scientific notation $n \times 10^x$, all digits comprising n are significant by the first six rules; 10 and x are not significant.* The value 5.02×10^4 has three significant figures. Rule 7 provides an opportunity to change the number of significant figures in a value by manipulating its form. For example, how do we write 1,100 with three significant figures? Given rule 6, 1,100 has two significant figures; its two trailing zeros are not significant. If we add a decimal to the end, we have "1,100." with four significant figures (by rule 5). But by writing it in scientific notation (1.10×10^3), we create a three-significant-figure value.

The precision of a number is represented by the number of significant figures it contains. A simple way of determining the number of significant figures is by writing the number in scientific notation (see note 8 in chapter 3)—1,000 = 1 × 10³, 1000.1 = 1.0001 × 10³, and 0.0045 = 4.5 × 10⁻³—and then counting the number of digits in the prefix (one, five, and two in these examples).

An important rule when calculating a number derived from several measurements requires one to report the result with the number of significant digits of the *least* precise measurement plus one. If you use fewer digits than this, you throw away part of the information you have received. If you use more digits, you are "inventing" information that you do not actually have.

For example, take the ratio 2/3. It is mathematically correct to approximate this fraction as 0.66666666667. But if the numbers 2 and 3 are obtained from measurements with limited precision, you must report their calculated ratio more conservatively. Say a scientist records the speed of a moving object as covering 2 meters in 3 seconds. The correct quotation for the velocity of the object is *not* 0.66666666667 but rather 0.67 m/s. The expression 0.66666666667 m/s has a precision of eleven decimal places and implies that the scientist knows the speed to one part in a trillion. That would be quite a high-tech experiment indeed! In fact, "2 meters" and "3 seconds" each have a precision of only one significant figure. Following our convention, the answer must have two significant figures, where the appropriate rounding (see box 7.4) has occurred.[7] We will attain the same result if the data were reported as 2 meters in 3.005 seconds. The expression 2 meters, which has one significant figure, denotes our least precise measurement (compared with 3.005 seconds, which has four significant figures)—and the least *precise* number determines the *accuracy* of the result and therefore the *precision* to which it should be recorded.

It is ironic that despite rampant innumeracy among its practitioners, the news media are obsessed with precise (but not necessarily accurate) numbers. As but one example (others may be found throughout the

BOX 7.4 THE CORRECT METHOD OF ROUNDING NUMBERS

With the exception of enumerating discrete objects (e.g., 5 oranges, 113 orangutans), a measurement is always an approximation. For example, a friend asks you what time it is. You look at your digital phone display and see that it says 2:44:35. Are you going to tell your friend it's 2:44 and 35 seconds? I suspect not. You'll probably be thoughtful enough to approximate by rounding up to 2:45. If, on the other hand, it's 2:32:14, you might round down to 2:30.

In mathematics, rounding has a more formal definition. The final digit (or significant figure) of any number is actually an approximation. To round off a number to n significant figures, the following rules apply:

1. If the digit to the right of the last digit you want to keep (that is, the first digit you want to drop off, $n + 1$) is less than five, then drop it (and everything else to its right). The value left behind is your rounded value.
2. If the digit in the $n + 1$ place is greater than five then drop it (and everything to its right) and raise the last remaining digit by one.
3. If the digit in the $n + 1$ place is equal to five, drop it if the preceding (nth) digit is even—leave n alone; if the nth digit is odd raise it by one. This convention is necessary to keep a set of rounded numbers as "balanced" as possible; i.e., if you round down for digits one through five (five cases) and up for six through nine (only four cases), the sum of the resulting numbers will be, on average, lower than the sum of the unrounded terms. If you run a bank, you could make money by adopting this scheme in paying interest to your customers, but it would violate a commitment to good scientific habits of mind.

See note 7 for examples illustrating these rules.

book), I recount an incident relayed to me by a colleague, Professor Nick Suntzeff, now the director of the Astronomy program at Texas A&M, but in 2010–2011 a Jefferson Fellow at the U.S. Department of State. The Jefferson Fellowship is administered by the U.S. National Academies of Sciences and Engineering to bring some scientific reasoning and knowledge to our foreign policy apparatus.

During the overthrow of Muammar Gaddafi in Libya, the situation on the ground became chaotic and foreign nationals were being evacuated. One afternoon, Nick was taking his turn answering the phone at the State Department and received a call from CNN wanting to know how many U.S. citizens were trying to get out of Libya. His estimate, consistent with the data the department had at the time, was 160 ± 10. The caller couldn't handle the notion of uncertainty and, after repeating the question of *exactly* how many U.S. nationals were in Libya and not getting what he wanted, he hung up. The next day, hoping to get someone else, the CNN researcher called back. Again he got Nick. This time Nick tried to explain what plus or minus meant and finally said between 150 and 170. The caller insisted that was not good enough. Nick finally gave in: 166. Within ten minutes CNN anchor Wolf Blitzer intoned to audiences around the world: "We now know there are 166 U.S. citizens trying to get out of Libya."

I guess it is comforting for people to have certainty even when it is meaningless.

ERROR AND UNCERTAINTY

All measurements (with the possible exception of counting discrete items) have associated uncertainties. Adopting a somewhat sloppy use of English, scientists often refer to such uncertainties as errors. They are not errors in the sense of mistakes. Rather they represent an estimate of the amount by which a measurement may differ from the true value of the measured quantity. In the quantum realm of individual atoms, there is an inherent uncertainty in measurement that cannot be overcome by clever instruments or more careful procedures. In the macroscopic world, however, measurement error inevitably results from using instruments of finite precision that must accomplish a measuring task in a finite amount of time.

If you measure the length of your desk with a foot-long ruler as accurately as you can and then ask your neighbor to bring over his ruler and do the same, it is quite unlikely that you will get exactly the same answer. Two types of errors will lead to the discrepant results: systematic errors and statistical (or random) errors.

Systematic errors arise from such problems as the fact that the two rulers will not be precisely the same length, or that the temperature of the room may change between the measurements, and the desk may have expanded or contracted a bit as a result. There may also be differences in the way that you and your neighbor approach the task. In some cases, sources of systematic error can be reduced or eliminated: e.g., you could make sure the room temperature remained constant or ship the desk to Sèvres outside of Paris where the world's standard meter stick resides and use that to measure the length. Often, however, we are unaware of systematic errors—"How was I supposed to know the cheap bookstore rulers are really only 11.9 inches long?"—or have no easy way of controlling them (keeping the temperature and all other potential influences perfectly constant). Repeating a measurement using different instruments—or, even better, different techniques—is the best way to discover and correct systematic errors.

Random errors may also be difficult to eliminate, but they are both quantifiable and reducible, and the subject of statistics holds the key. In the case of measuring the length of the desk, random errors arise from such things as not perfectly aligning the ruler with the edge of the desk or not precisely moving the ruler exactly one ruler's length each time. Unlike systematic errors such as those produced by the short ruler, random errors—as their name suggests—have both plusses and minuses, producing answers too long and too short with roughly equal frequency. As I describe in the next section, the random nature of these errors make them more tractable.

STATISTICAL DISTRIBUTIONS

THE GAUSSIAN DISTRIBUTION

Any good scientific measurement must be qualified by stating an uncertainty, or error. It is standard practice to notate this by quoting a result plus or minus an error. In some cases, random and systematic errors are reported consecutively; e.g., 5.02 ± 0.17 ± 0.2 ft. translates to a random error of 0.17 ft. and a systematic error of 0.2 ft. What does the ± sign mean? Focusing on the random error for the moment, it does not, as one might expect, imply that the true length of the desk is definitely between 4.85 and 5.19 ft. Rather, it represents shorthand for describing the *probability* that the quoted measurement is within a certain range. Although the distribution of errors, and thus the range of probabilities, is not always easy to determine, in many cases of interest it is safe to assume that the individual measurements are distributed in a "normal distribution," more commonly referred to as a bell curve. This curve is also known as a Gaussian curve after the mathematician Karl Gauss. It is described by the equation $P(x) = [\sigma \sqrt{2\pi}]^{-1} \times e^{-(x - \mu)^2/2\sigma^2}$, where σ = standard deviation and μ = mean of x.

This expression is plotted in figure 7.2. The curve's peak represents the average or mean value of all the measurements: if you had the time to make an infinite number of measurements so that all the random errors averaged out, this would represent the true value of the quantity of interest (e.g., the length of the desk). The value of sigma (σ) determines the width of the curve and is called the standard deviation.[8]

It is logical to ask at this point how we can know the value of σ. In other words, how do we know the width of the distribution of errors in our measurement? The most straightforward way to determine this quantity is actually to measure it. That is, if you perform the desk-length measuring exercise 100 times and plot the results in the form of a histogram (see chapter 5), it will probably look very much like figure 7.3. As you can see, this is approximated quite well by a Gaussian distribution.

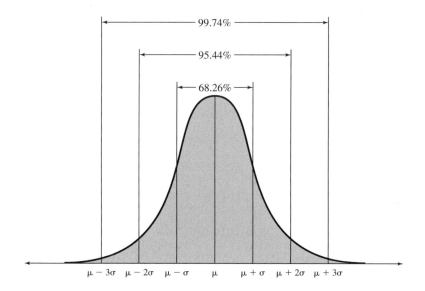

FIGURE 7.2 THE GAUSSIAN DISTRIBUTION

A Gaussian distribution in which the symbol mu (μ) represents the mean of the distribution and sigma (σ) is one standard deviation. The fractions of the curve enclosed by +/- 1, 2, and 3 sigma are shown.

The standard deviation is a convenient way of characterizing the spread of a set of measurements. But if you *do* go through the trouble of repeating a measurement many times, shouldn't you gain something beyond just knowing how big your random errors are? Shouldn't doing many measurements improve the accuracy of the answer? Indeed it does. The quantity σ will not change if you take twenty measurements or 200—it is just a description of the irreducible errors associated with your measuring technique. But random errors, by their very nature, tend to average out if many measurements are summed. The average (or mean) value of twenty-five estimates is a better approximation of the true value than an average of nine. The "error" quoted on such mean values, then, is not the error that characterizes each

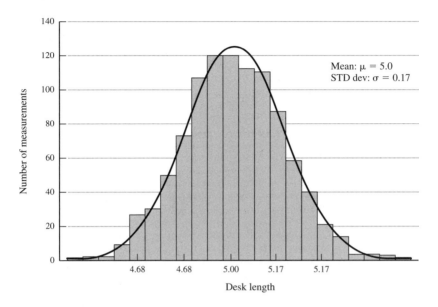

FIGURE 7.3 A THOUSAND MEASUREMENTS APPROXIMATES A GAUSSIAN

An imaginary set of measurements of desk length plotted as a histogram. The shape of the overall distribution looks very much like the Gaussian curve in figure 7.2, which is plotted over the histogram shown here.

measurement but the error in the mean. It turns out this is simply given by $\sigma/\sqrt{(n)}$, where n is the number of measurements you made. Thus, to be quantitative, twenty-five measurements are $5 = \sqrt{(25)}$ times better than one.

As a practical example, we can apply this concept to our calculation of the mean lunar phase during the Son of Sam shootings. We already calculated the mean of the times between the full moon and when the shootings occurred. Using note 8, we can calculate the standard deviation of these values as 5.29 days and the error in the mean (with eight measurements) $5.29/\sqrt{8} = 1.9$ days. Thus, the observed mean value of 8.24 days is fully consistent with the expectation of 7.36 days if the Moon has no effect to within less than one-half of a standard deviation and is inconsistent with the lunatic expectation of 0.0 at the 99.99 percent confidence level (almost four standard deviations).

The usual practice in quoting the statistical error of a measurement is to quote plus or minus one sigma, that is plus or minus one standard deviation. Integrating under the Gaussian curve reveals that 68 percent of its area lies within plus or minus one sigma of the mean. Thus, the literal meaning of the measurement of the desk's length reported as 5.02 ± 0.17 ft. is that the measurement is consistent with the true length of the desk lying in the range 4.85 to 5.19 ft. at the one-sigma level (meaning there is only a 32 percent chance that the true value lies outside this range). If you have done the measurement twenty-five times, however, you might choose to advertise your diligence by instead quoting the error in the mean or $5.02 ± 0.17/\sqrt{(25)} = 5.02 ± 0.034$, which implies that the true value has a probability of 67 percent of lying between 5.054 and 4.986 ft. If you want to be more conservative about your knowledge of the length of the desk, you can use a two-sigma uncertainty—$5.02 ± 0.068$ (2σ), which, owing to the shape of the Gaussian curve, means the true value has a 95 percent chance of lying between 4.952 and 5.088 ft.

THE BINOMIAL DISTRIBUTION

The Gaussian distribution allows for errors of any size; i.e., it is a continuous distribution as is appropriate for, say, measuring the length of a desk that might be in error by 0.1237 or 0.1238 inches. In many cases, however, a finite number of measurements may yield only one of a discrete number of outcomes.

Suppose you were to do the experiment of flipping a coin five times. The result of each experiment is unambiguous: it is either heads or tails. There is no "error" in each outcome. However, you might be interested in knowing how many times when you do this experiment you are likely to get three heads. Harkening back to our definition of probability, all this requires is to enumerate all the possible outcomes of the experiment and then divide that number into the number of outcomes of interest—in this case three heads. With only five flips, this is a practical, if tedious, approach; the possibilities are enumerated (and the probability calculated) in box 7.5. But suppose you wanted to know how many times fifty flips would yield thirty-one heads—or, equivalently, how many times you would catch thirty-one female frogs out of fifty total frogs caught if the gender ratio in a pond were actually 50:50. It would clearly be useful to have a general formula to calculate the probabilities of these outcomes.

And there is one. It is called the binomial distribution. The derivation is not overly complicated, but I will spare you the details here. The probability of m successes in n trials when the probability of the outcome of interest is P is given by $P(m$ successes in n trials$) = n!/[m!(n - m)!] \times P^m \times (1 - P)^{(n - m)}$, where $n!$ is just the factorial that equals $1 \times 2 \times 3 \times 4 \times (n - 1) \times n$.

Okay, so it looks complicated, but it is not *nearly* as bad as writing down every possibility for catching fifty frogs in a putatively gender-neutral pond and then counting up how many of them give you thirty-one females. For example, let's use this distribution to determine the chance of getting three heads in five flips: $n = 5$, $m = 3$, and $P = 1/2$ (the odds of getting heads is 50 percent), so $5!/[3!(2!)] \times (1/2)^3 \times (1/2)^2 = 10 \times (1/8) \times (1/4) = 0.31$ or 31 percent, which equals 10/32, the answer we derived from the tedious enumeration in box 7.5.

For the frogs, the answer is 0.027—only 2.7 percent. If you got this result, you might want to abandon your hypothesis that the gender ratio in your frog pond is really 50:50 and search for an explanation for the preponderance of females.

THE POISSON DISTRIBUTION

In some experiments, the goal is simply to count events that are occurring at random times and compare this to the known average rate at which

BOX 7.5 AN ILLUSTRATION OF THE POWER OF
THE BINOMIAL DISTRIBUTION

The number of possible outcomes of each coin flip is two (either heads or tails). So the probability of either a head or a tail is 1/2. In chapter 6 we described how the number of possible outcomes of several independent events is the product of the number of possible outcomes of each event individually. So the number of combinations that two coin flips will give you is $2 \times 2 = 4$. In this case we are flipping five coins, so the number of possibilities is $2 \times 2 \times 2 \times 2 \times 2 = 32$. One way to answer the question of how many times you should expect to get three heads in five flips is to list all possible combinations and count the outcomes:

```
 1  h h h h h (5 h)
 2  h h h h t (4 h)
 3  h h h t h
 4  h h t h h
 5  h t h h h
 6  t h h h h
 7  h h h t t (3 h)
 8  h h t h t
 9  h t h h t
10  t h h h t
11  h h t t h
12  h t h t h
13  t h h t h
14  h t t h h
15  t h t h h
16  t t h h h
17  t t t h h (2 h)
18  t t h t h
19  t h t t h
20  h t t t h
21  t t h h t
22  t h t h t
23  h t t h t
24  t h h t t
25  h t h t t
26  h h t t t
27  t t t t h (1 h)
28  t t t h t
29  t t h t t
30  t h t t t
31  h t t t t
32  t t t t t (0 h)
```

As you can see for yourself, there are ten possible ways to get three heads. Thus, the probability of getting three heads from five coin flips is 10/32 or 5/16. The binomial distribution allows you to simply calculate this result without the tedium of listing every possible outcome.

they are expected to occur. One of the important examples of this is in using the radioactive decay of atomic nuclei as clocks to date ancient materials. We now know the age of the Earth to better than 1 percent through the application of such techniques and have dated the Shroud of Turin (the putative burial cloth of Jesus) to the 1350s (when church documents show it was forged by an artist in the employment of a corrupt bishop). In these applications, we use the fact that each type of radioactive nucleus decays at a precisely determined average rate, although the individual decays happen at random moments. For example, the heavy isotope (see box 10.5) of carbon-14 (C-14) will have half of its atoms decay over an interval of 5,730 years. We can determine how much of the C-14 breathed in by a plant (or the linen made from it) is left by counting the number of decays (signaled by the emission of a high-energy electron) and thus determine the age of the sample directly.

The uncertainty in this kind of counting experiment is given by a third kind of distribution: the Poisson distribution. For an observed number of counts n and an average number expected a, the probability of getting n counts in a given trial is $P_a(n) = (a^n e^{-a})/n!$

For example, if an archeologist found a fragment of cloth in a Central American excavation and wanted to know its age, she might turn to C-14 dating. The cloth is of a relatively tight weave. Did the first New World settlers 15,000 years ago know how to weave so skillfully (did aliens teach them how to do it?), or is the fragment from the time of the conquistadors, only 500 years ago? Putting the cloth in a device to count radioactive decays, we record one, then two, then zero, then zero, and then two counts in the first five one-minute counting intervals. Given the size of the sample and the decay rate of C-14, we would expect an average counting rate of 0.917 decays per minute if the cloth is 500 years old and only 0.074 decays per minute if it is 15,000 years old. The Poisson distribution tells us the probability of getting zero, one, and two counts in each one-minute interval. In this case, we expect zero counts 40 percent of the time, one count 37 percent of the time, and two counts 17 percent of the time if the cloth is 500 years old and zero counts 93 percent of the time, one count 6.8 percent of the time, and two counts 0.2 percent of the time if it is 15,000 years old. Clearly, our results favor the younger age—there is only a 1 in 500 chance we would see two counts in a single minute, whereas this happened in the first five minutes. Continuing the experiment for another ten minutes or so would seal the case.

In fact, the Poisson and Gaussian distributions are both special cases of the binomial distribution. We use the former in situations where the expected events are relatively rare and their average occurrence rate is well-known. The Gaussian distribution is useful when our sample size is large. In all cases, however, these probability distributions allow us to dispassionately assess the outcome of an experiment or an observation and help guide our future hypotheses. The result of the Central American cloth dating suggests it would be a waste of time to pursue the alien weaver hypothesis, whereas the frog gender ratio might well lead the ecologist to question whether polluting chemicals might be altering the natural ratio of males to females and prompt him to propose further experiments.

And this is the real purpose of statistics: to provide the scientist with a quantitative assessment of the uncertainty in a measurement and the likelihood that, given an assumed model, the measured value is consistent with the model's prediction. Note that I say "consistent with" and not that a measurement *proves* that a model is correct. As noted earlier, the world of science is not about proofs (that's the realm of mathematics and philosophy). Although often a precise discipline that strives for accuracy, science is ever-aware of the inherent and unavoidable uncertainties in its measurements and accounts for them explicitly when building and testing models of the natural world through the enlightened application of statistics.

Statistics aren't lies at all, but they can be very useful in uncovering them, adding a layer of protection against the assaults of the Misinformation Age.

8
CORRELATION, CAUSATION . . . CONFUSION AND CLARITY

t was finally time to give up on the drugstore glasses and get a real prescription so I could see my computer screen without squinting. I hadn't been to an ophthalmologist in some years, and my wife had a new one she liked near our house on the North Fork of Long Island, so I agreed to make an appointment.

Since I just *knew* what he was going to say when he looked into my left eye, I said, "Before you look, let me warn you that my pattern of capillaries is unusual. My first eye doctor told me about it forty years ago."

As predicted, he looked, paused, and then said slowly, "Yes. That's concerning, as that pattern is highly correlated with brain tumors."

And I wasn't the least alarmed. Why? Because (1) as I had just told the doctor, the pattern had been there for at least forty years, probably since birth, and (2) I understand what "correlation" means—and what it doesn't.

We have all seen headlines such as the following:

Smoke and Dust at World Trade Center Are Linked to Smaller Babies

Belief in Hell Brings Prosperity to Nations

> Vegetarianism Increases Life Span
> Premature Babies Have Smaller Brains in Youth
> TV Watching by Toddlers Leads to Attention Deficit Disorder
> Five Hours of TV a Day Makes Kids Six Times More Likely to Smoke

I culled all of these claims not from headlines in *The National Enquirer*, but from the *New York Times*. Some of these headlines are "common knowledge," some sound plausible, and at least one sounds completely silly. All of these claims are based on one of the most widely misused and misunderstood mathematical arrows in the scientist's quiver: correlation analysis. This chapter aims to demystify this relatively simple technique and to provide you with some ground rules for assessing any claims based on correlations lest you fall prey to the traps laid for you in the Misinformation Age.

I begin with one of my favorite examples: the correlation between air pollution and premature death. Several decades ago in western Pennsylvania, a study collected the records of both air pollution and deaths due to pulmonary disease from several counties around Pittsburgh when that city was the nation's leading center of steel production. The authors reported a correlation between poor air quality and the death rate. In particular, Allegheny County had the worst air and the greatest number of deaths.

The conclusion seems obvious: the correlation between air pollution and pulmonary disease implies that air pollution causes deaths. Pollution control laws were just being implemented at the time, yet, as one wag pointed out, there were a few sites exempted by the county's laws: crematoria. Perhaps, then, the study's authors had their conclusions backward; really, it is deaths that cause air pollution.

Several points about this story are worth noting. Although the true relation between air pollution and pulmonary disease may seem obvious, a correlation between two varying quantities can never be taken as prima facie evidence that one causes the other. A co-(r)relation is just that: a relationship between two quantities that vary together. It is not a statement about cause and effect. Viewed in isolation, it is not possible to tell what the relationship between two correlated variables is: A could cause B, B could cause A, or a third (or fourth or fifth) factor could cause both A and B.

I hear you thinking: "Pedant! It's common sense that air pollution causes pulmonary disease. Why do we need to worry about statistics and uncertainties—the cause is obvious."

Einstein's cautionary words about common sense—that it is the layer of prejudices laid down upon the mind prior to the age of eighteen—bears repeating here. Since most of my readers are well past eighteen, it is time to root out and unlearn some of those prejudices. Indeed, this is essential if one is to develop the scientific habits of mind required to survive the Misinformation Age.

CORRELATION DEFINED

As noted in chapter 5, the scatter plot—the display of pairs of measurements on a two-dimensional graph—is a useful tool for finding patterns in data. But as that chapter also made clear, humans are good at finding patterns even when none exist. What we require is an objective, quantitative measure of whether the patterns we see are significant. For example, it appeared from our plot in chapter 5 that the stronger the belief that special, innate talent was required for success in a field, the fewer women received PhDs in that field (see figure 8.1).

Our goal now is to quantify this apparent effect and to determine its significance; i.e., is it a chance effect that arises from our relatively small samples (some disciplines only had a couple of dozen respondents) or one worth examining further to seek out mechanisms for how such beliefs translate into an underrepresentation of women in certain fields?

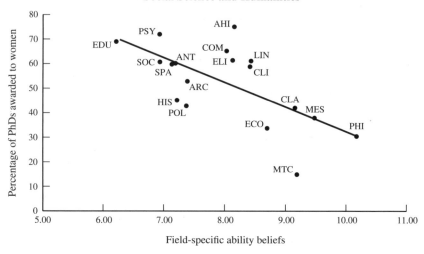

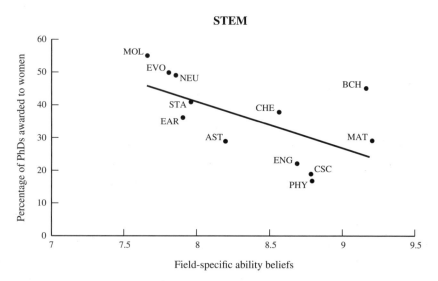

FIGURE 8.1 WOMEN PHD PRODUCTION AND "FIELD-SPECIFIC ABILITY BELIEFS"

Plot comparing field-specific ability beliefs with the fraction of women receiving PhDs in various academic disciplines as discussed in chapter 5.

To assess the relationship between pairs of variables, we calculate the "linear correlation coefficient," often called simply the "correlation coefficient," which is denoted by the symbol r. The correlation coefficient is a quantity, calculable from the data themselves, that signifies how closely the data points approximate a straight line of the familiar form $y = mx + b$. It is defined as follows:

$$r = \frac{\Sigma \left(x_i - \langle x \rangle \right)\left(y_i - \langle y \rangle \right)}{\left[\Sigma \left(x_i - \langle x \rangle \right)^2 \Sigma \left(y_i - \langle y \rangle \right)^2 \right]^{1/2}}$$

It is relatively straightforward (although pretty boring) to prove (in the mathematician's sense—it is really true) that r can take on any value between +1.0 and –1.0. A value of +1.0 implies a perfect correlation; i.e., the data points all fall precisely on a line defined by the equation for a straight line $y = mx + b$. The value $r = -1.0$ means the data display a perfect inverse correlation or anticorrelation—the slope of the line is negative (i.e., if x gets bigger y gets smaller, and vice versa). If the data have no relationship whatsoever, they are said to be uncorrelated, and $r = 0$. Real data usually have some associated uncertainties; as a result, even in the case of a perfect underlying correlation, the results of an experiment will not likely yield an r value of exactly 1 or –1. Likewise, we never expect to find $r = 0.00$ precisely, even if the data have no correlation whatsoever.

CORRELATION APPLIED

Let's look at the data from figure 8.1 and calculate the value of r. First we find x, the mean of the "ability beliefs" variable, and y, the average percentage of women receiving PhDs in the thirty academic fields studied, by summing each set of numbers and dividing by thirty. We then calculate the differences of each point from the mean and sum them (squared or not) as specified in the equation given previously. The result is that $r = -0.60$, which strongly suggests a negative (or anti-) correlation.

Now that we have a value for r, what do we conclude? Is the trend "significant" in the sense that we discussed this word in chapter 7 and expand upon in this chapter? What is the probability that we would find such a value for r by chance if, in fact, there were no actual relationship between the two variables? This is clearly a statistical question, and although a rather complicated probability to assess, it can be done, and the results have been collected in a convenient tabular form. Having calculated r for the total number of data points, n, we can simply look up the probability $P_n(|r| \geq r_o)$, which is the probability, given n measurements, that we would find an absolute value of r bigger than the listed r_o if no real correlation existed. We use the absolute value sign since large values of r can be significant in either the positive or negative sense—the data could be either highly correlated or highly anticorrelated, with r near 1 or –1. The result is provided in table 8.1.

Using the data from all disciplines, $n = 30$, and the correlation coefficient has a value $r = -0.60$. Looking at table 8.1 shows that the probability of this value arising by chance is <0.001. If we use only the social science/humanities data, $n = 20$ and $r = -0.62$, so the probability of a chance occurrence of this level of anticorrelation is about 0.5 percent (note 0.62 itself does not appear in the table, so an extrapolation scheme[1] is needed to get the precise answer).

Different branches of science have different sociologically defined thresholds for what they consider significant. Most branches that deal with people (biology, medicine, psychology, neuroscience, etc.) accept two-sigma or 95 percent confidence as the level of significance worth taking seriously. With this criterion, we would certainly say that there is a significant anticorrelation between belief in the need for innate ability and the fraction of PhDs awarded to women in a field, since the probability of obtaining the result we did by chance is <0.1 percent for all fields

TABLE 8.1 PROBABILITY (%) THAT A CORRELATION WITH r_0 INVOLVING n DATA POINTS WILL ARISE BY CHANCE

	r_0										
n	0	0.1	0.2	0.3	0.4	0.5	0.6	0.7	0.8	0.9	1
3	100	94	87	81	74	67	59	51	41	29	0
4	100	90	80	70	60	50	40	30	20	10	0
5	100	87	75	62	50	39	28	19	10	3.7	0
6	100	85	70	56	43	31	21	12	5.6	1.4	0
7	100	83	67	51	37	25	15	8.0	3.1	0.6	0
8	100	81	63	47	33	21	12	5.3	1.7	0.2	0
9	100	80	61	43	29	17	8.8	3.6	1.0	0.1	0
10	100	78	58	43	25	14	6.7	2.4	0.5	—	0
11	100	77	56	37	22	12	5.1	1.6	0.3	—	0
12	100	76	53	34	20	9.8	3.9	1.1	0.2	—	0
13	100	75	51	32	18	8.2	3.0	0.8	0.1	—	0
14	100	73	49	30	16	6.9	2.3	0.5	0.1	—	0
15	100	72	47	28	14	5.8	1.8	0.4	—	—	0
16	100	71	46	26	12	4.9	1.4	0.3	—	—	0
17	100	70	44	24	11	4.1	1.1	0.2	—	—	0
18	100	69	43	23	10	3.5	0.8	0.1	—	—	0
19	100	68	41	21	9.0	2.9	0.7	0.1	—	—	0
20	100	67	40	20	8.1	2.5	0.5	0.1	—	—	0
25	100	63	34	15	4.8	1.1	0.2	—	—	—	0
30	100	60	29	11	2.9	0.5	—	—	—	—	0
35	100	57	25	8.0	1.7	0.2	—	—	—	—	0
40	100	54	22	6.0	1.1	0.1	—	—	—	—	0
45	100	51	19	4.5	0.6	—	—	—	—	—	0
	0	0.05	0.1	0.15	0.2	0.25	0.3	0.35	0.4	0.45	
50	100	73	49	30	16	8.0	3.4	1.3	0.4	0.1	
60	100	70	45	25	13	5.4	2.0	0.6	0.2	—	
70	100	68	41	22	9.7	3.7	1.2	0.3	0.1	—	
80	100	66	38	18	7.5	2.5	0.7	0.1	—	—	
90	100	64	35	18	7.5	2.5	0.7	0.1	—	—	
100	100	62	32	14	4.6	1.2	0.2	—	—	—	

Note the substantial change in value of r_0 in the rows following n = 45.

and ~0.6 percent for the humanities and social sciences. Physical scientists tend to adopt a three-sigma threshold (99.7 percent likely, allowing only a 0.3 percent probability of chance occurrence). Our result for the sample as a whole even meets this more stringent test; table 8.1 does not even bother to show a value in the $n = 30$, $r = 0.6$ location because the result is significant at more than the 99.9 percent level and thus would be accepted as meaningful by any good scientist, although the social sciences/humanities relations would not quite meet this higher threshold.

Thus, our analysis has shown that there is a significant anticorrelation between these two variables. Note that this statement—which is all we can fairly draw from our analysis—provides no information whatsoever about the cause of this relationship. That we must seek through further observation or experimentation.

A second example of correlation analysis may be of interest to university students and to their parents. Some years ago, I collected anonymous data for 100 randomly selected members of a Columbia College class that included verbal and math SAT scores plus GPAs at the end of their first year at Columbia. The mean values with their associated statistical uncertainties (quoted as errors in the mean—see chapter 7) were as follows: verbal SAT = 691.5 ± 7.6, math SAT = 695.7 ± 6.8, and first-year GPA = 3.374 ± 0.051. These values were consistent with the entire class averages (see box 8.1).

The value for the correlation coefficients were r(verbal GPA) = 0.547 and r(math GPA) = 0.436; for $n = 100$, table 8.1 shows both are highly significant. For a sample of this size, a value of $r > 0.35$ will occur by chance less than twice in 1,000 trials; values of ~0.4 and ~0.5 are much less likely still. We can clearly say that the correlation is significant.

Can I use this result to predict any individual student's performance before he or she has even taken a single Columbia exam? Should we just grant those students with high SAT scores their degrees when they enter the place and save a lot of time and energy?

BOX 8.1 THE MEANING OF CONSISTENCY

For the entire class, the actual mean verbal and math SAT scores were 703 and 704, respectively. I say that the values I derived by using only 10 percent of the class were consistent with these global values because the mean values for our subsample—verbal SAT = 691.5 ± 7.6 and math SAT 695.7 ± 6.8—were the same within 1.5 and 1.2 standard deviations, respectively. That is, there was no statistically significant difference between this 10 percent of the class and the whole class. If I had made up 100 different subsamples, each containing 100 students (or 10 percent of the class), I would find 13 more than 1.5 sigma away from the overall mean and 23 more than 1.2 sigma off. Thus, the discrepancies of 12.5 and 8.3 points in these values are just the normal, random statistical fluctuations one should expect when looking at only 10 percent of a 1,000-student class.

Clearly not. The data are shown respectively in figures 8.2 and 8.3. Several points are worth noting. First of all, with a sample size of 100, most of the possibilities are represented, and the diagrams are sprinkled almost everywhere with points within the allowed ranges; e.g., SAT scores are always less than or equal to 800. This makes the correlation, although highly statistically significant, less apparent to the eye than in the example on women's PhDs discussed previously that had only one-fifth as much data. A corollary of this fact is that there are students with SAT scores in the low 500s who earn A averages and students with 750 SATs who have year-end GPAs a full grade or more lower. Even students with 800 SAT scores can find themselves with a B average after their first year. Furthermore, since we have sampled only 10 percent of the class, the dataset is unlikely to include the most extreme outliers; the combination of perfect

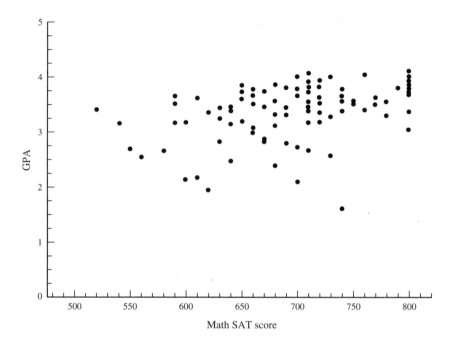

FIGURE 8.2 MATH SAT SCORES AND STUDENT GPAS

Math SAT scores versus first-year grade point average (GPA) for a sample of Columbia College students.

SAT scores and C averages, as well as 500 SAT scores and A averages, are not at all unprecedented.

This illustrates a crucially important idea: *a correlation is not predictive for individual cases*. It is strictly a statistical statement about how two variables are related in aggregate. Just as correlation is not causation, correlation is not destiny. Just as many different factors contributed to a student's final SAT scores, many different and new factors will contribute to his or her success in college.

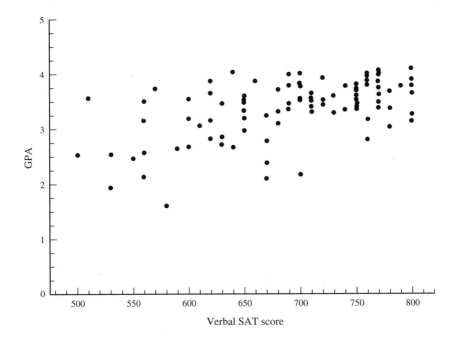

FIGURE 8.3   VERBAL SAT SCORES AND STUDENT GPAS

Verbal SAT scores versus first-year grade point average (GPA) for the same sample as in figure 8.2.

CORRELATIONS MISAPPLIED

The definition of the linear correlation described previously is mathematically precise. It is critically important, however, not to imbue this precision with meaning. Perhaps the best, and certainly the most hilarious, illustration of the meaninglessness of many correlations can be found on Tyler Vigen's website www.tylervigen.com. Some of his most compelling findings are as follows:

1. Between 2000 and 2009, there was an extremely strong correlation (r = 0.95) between per capita cheese consumption in the United States and the number of people who died by becoming tangled in their bed sheets (which, remarkably, peaked in 2008 at over 800 people per year).

2. In that same period, the divorce rate in Maine was almost perfectly correlated (r = 0.993) with the US per capita consumption of margarine (which, interestingly, fell from 8 to 3.7 lb. per person over this period).

3. The number of bee colonies in the United States between 1990 and 2009 showed a very strong anticorrelation with juvenile arrests for marijuana possession (r = -0.93). (Yes, I did Google "marijuana pollination" and while I found 114,000 helpful websites, none of the first fifty, at least, mentioned bees.)

While these and Vigen's many other outlandish examples are amusing—and completely spurious—they are also harmless: no one is going to take action based on these correlations (although I did hear a radio host once, when interviewing Vigen, say he was going to stop eating cheese before bedtime). Unfortunately, as discussed in chapter 12, many clinical studies in medical fields report equally meaningless correlations. The problem is that these reports are frequently picked up by the media and trumpeted as important new information. The result is both a propagation of misinformation and, when not confirmed by follow-up studies, another reason for the audience to distrust "science."

Finally, in case you were wondering, I don't have a brain tumor. My odd capillary pattern may *correlate* with tumors, but it doesn't *cause* such tumors. Correlation analysis can be a powerful tool in scientific analysis, but applied unthoughtfully, it can also add to the immense amount of nonsense that fuels the Misinformation Age.

9
DEFINITIONAL FEATURES OF SCIENCE

"Is there any way to prove your theory is false?"
. . . (*long pause*) "No."

Y ES! After eight straight hours in a windowless conference room, that "No" made the whole day worthwhile.

It was 1981—the interregnum in Bill Clinton's governorship of Arkansas.[1] The state legislature had passed, and the new governor Frank D. White had signed Act 590, which required the teaching of creationism in all Arkansas science classrooms. This was the first major victory in eleven years for the "creation science" movement led by such figures as Henry M. Morris and Martin E. Clark of the Creation Research Society,[2] whose views are best articulated in the following quote from their book *The Bible Has the Answer*:

> Evolution is thus not only anti-Biblical and anti-Christian, but it is utterly unscientific and impossible as well. But it has served effectively as the pseudo-scientific basis of atheism, agnosticism, socialism, fascism, and numerous other false and dangerous philosophies over the past century.[3]

Fortunately, not all Arkansans were delighted with this turn of events. The leaders of all the state's mainstream churches, teachers, parents, and national organizations (religious and otherwise) joined in a federal lawsuit seeking to block the implementation of Act 590. The suit was not an attack on religion or even on the place of religion in public schools. The issue was what constituted science and thus belonged in science classrooms. The American Civil Liberties Union (ACLU) provided lawyers to prepare the case.

Unfortunately, very few lawyers have much of a background in science. Many of the expert witnesses called by the State from the Creation Science Research Center (CSRC) in San Diego and elsewhere, however, have PhDs in science. So the ACLU brought in reinforcements. I was asked to assist in deposing a witness[4] and arrived as requested at 9 A.M. on a high floor of one of those indistinguishable (and undistinguished) glass and steel towers on the East Side of Manhattan. After a brief conversation with the ACLU lawyer, I was ushered into the conference room to meet the Arkansas attorney general (AG), the court stenographer, the witness, and a PhD physicist and member of the CSRC. I was the only person not allowed to speak—all my questions and rebuttals had to be written on a pad and passed to the ACLU lawyer (even the stenographer had a machine!). It was clearly going to be a long day.

Long, but fascinating. As noted previously, the key issue in the case was the very definition of science, the subject one should expect to find in a science classroom. The "scientific creationists" had been working for more than a decade on wrapping the literal account of creation in the Bible with the trappings of science. They had scientific meetings, journals, theories, data, lots of jargon—everything that made their "research" sound like science to the uninitiated. The one crucial element they lacked, however, was the essential ingredient of any scientific theory: falsifiability. As described in chapter 2, for a model or theory to reside within the realm of science, it must, in principle at least, be possible to demonstrate that the model or theory is false—that is, incompatible with direct observations of Nature. Our strategy, then, was to trap the witness into admitting that there was no experiment that could be performed and no observation that could be made that could change his "model" of creation.

It took eight hours. The ultimate exchange went roughly as follows:

WITNESS: The short-period comets provide yet another piece of evidence inconsistent with the timescales required by the theory of evolution.

ACLU LAWYER: How so?

WITNESS: Each time a comet passes close to the Sun in its elliptical orbit, large amounts of material (ice and rocky debris) are blown away by the Sun's heat and light. We can calculate how long these comets can exist by knowing their orbit periods and measuring how much mass is lost with each solar passage. We find they are at most a few thousand years old. Since comets are widely acknowledged to be primordial solar system material—the detritus left over from the formation of the planets themselves—it is clear that the solar system cannot be more than roughly 10,000 years old, or there would be no short-period comets left.

The ACLU lawyer looked at me in panic—it sounded like an airtight argument to him. Indeed, all the facts adduced were fully consistent with our scientific understanding of comets—right up until the last phrase. I scrawled: Are comets affected by gravity?

ACLU LAWYER: Are comets affected by gravity?

WITNESS: Yes. That's what holds them . . .

INTERRUPTION BY ARKANSAS AG: You needn't elaborate, just yes or no will do.

ME (*STILL SCRAWLING*): By Jupiter's gravity?

ACLU LAWYER: Are they affected by Jupiter's gravity?

WITNESS: Yes.

ME: Can Jupiter's gravity change their orbit periods?

ACLU LAWYER: Can Jupiter change their orbit periods?

WITNESS: Yes.

ME: Can it change long-period comets into short-period ones?

ACLU LAWYER (*FINALLY FEELING MORE CONFIDENT*): And can Jupiter's gravity then change the orbit of a comet with a long period into one of these short-period comets?

WITNESS (*AFTER A PAUSE*): Some people think so.

ACLU LAWYER: Do you?

WITNESS: No.

ACLU LAWYER: Is there any way to disprove your theory that these comets are young?

WITNESS (*LONG PAUSE*): No.

The actual trial began in Little Rock on my birthday in 1981. The decision was handed down on January 5, 1982:

> Pursuant to the Court's Memorandum Opinion filed this date, the defendants and each of them and all their servants and employees are hereby permanently enjoined from implementing in any manner Act 590 of the Acts of Arkansas of 1981.[5]

Although this particular battle was won, the war remains far from over. Repackaging creation science as "intelligent design," the fundamentalists are hard at work in many states, packing school boards, censoring textbooks, and pushing laws that require equal time for their alternative theories of evolution. Although the Bible and many other books—religious, philosophical, and otherwise—do offer alternative views of the world and its creation, they do not offer a "theory" in the scientific sense of the word. What is this scientific sense? Indeed, what are the defining features that make science science?

TRUTH AND FALSIFIABILITY

As outlined in chapter 2, science is not a search for Truth but a mode of thought that seeks falsifiable models of nature. This is the key distinction between the creationist's worldview and my own. The creationist's model, as a result of his faith in it, cannot conceivably be false; changing his model is fundamentally inconsistent with what he believes to be true. As a scientist, I am required to admit the possibility that his idea may be true—it may be the case that the universe, including the solar system and Earth, was formed 6,000 years ago. An omniscient and all-powerful being could indeed arrange for the light from each of the 10^{22} stars in the universe to be on its way to Earth tonight purely for my amusement. He could certainly have layered the rocks of the Grand Canyon to look very old, and he could have set ticking all the radioactive nuclei on Earth so that their decay rates falsely give an origin four-and-a-half billion years ago. This is a possible model for creation.

Indeed, in the course of the creationist's deposition, we reached a point where this idea of prearrangement arose, and he invoked the "five-second-agoers" in its defense. The ACLU lawyer looked blankly at me, and I looked equally blankly back. It turns out, five-second-agoers believe that everything could have been arranged just five seconds ago: all your memories, your plans, your bruises—and those memories others have of you, shared precisely as they are, for all time—could be a setup, constructed about the time you began reading this sentence.

They may call it a model, perhaps a theory even. And I cannot prove it is false. In fact, nobody can prove it false, because, by its very nature, it is not a falsifiable model. Hence, it is not science.

I have used the simple words data, model, and theory throughout this book. These words are often used in different contexts by people with varied backgrounds and predilections. When applied by nonscientists, they are, in fact, often misused. For example, students struggling in my introductory astronomy course often say something such as "I really like the theoretical aspects, but I can't get the math." To a physical scientist, this is an oxymoronic statement: an astrophysical theory *is* a mathematical formulation of an idea—the theory *is* the math (although this is not always the case in other branches of science).

To facilitate communication, definitions are useful. I provide here one scientist's view as to the meaning of these terms along with a few others that I believe are essential for adopting scientific habits of mind.

DATA

Contrary to the views of some postmodern literary theorists, scientists generally accept the notion that an objective reality exists and that one can sample that reality with measurements and observations that produce "data." In most branches of science, data are quantified; i.e., they are represented by numbers. These numbers can arise by simply counting—the number of tree species in your local park or the number of stars belonging to the Pleiades star cluster—or from some other process that yields a numerical value: the speed of nerve signal transmission or the depth of a glacial ice core. In the latter two cases, the numbers will have associated units that describe the quantities measured and tie them to a standardized system of measurement. The rate at which a neuronal signal travels will most likely be in meters per second, and glacial depth will be in meters or kilometers. Furthermore, all good measurements include an uncertainty or "error" (see chapter 7).

Data are the raw materials of science. They provide the basis for constructing models and the means by which such models are tested. They are gathered with great care. Every effort is made to minimize effects that might conflate the measurement of one quantity with another. Systematic errors are ruthlessly sought out and suppressed. Random errors are diminished through repeated measurements (see chapter 7). The great machines of science—telescopes, functional magnetic resonance imaging machines, particle colliders, oceanographic ships—are all designed to collect data with minimal interference from effects extraneous to the question at hand: what fact can I uncover about the universe today? Science divorced from data is not science.

PROXIES

Although quantitative data lie at the heart of our exploration of Nature, we are not always able to measure the thing we really want to know. Sometimes this results from the characteristics of the phenomenon under study and/or the limitations of our measuring instruments; it is difficult to imagine, for example, how we will ever directly observe elementary particles that pop in and out of existence in 10^{-23} seconds. In other circumstances, the numbers we really need simply do not exist; e.g., in charting the long-term history of Earth's climate, it is important to know the temperature of the planet over very long periods stretching back far before humans with their thermometers were around to record the numbers. In these and many other cases we resort to using "proxies," stand-ins for the true quantities of interest.

Take the example of global climate history. Paleoclimatologists (from *paleo*, meaning ancient and climatology, the study of climate) have developed highly reliable proxies for temperature in prehistoric times. One such indicator is the ratio of heavy- to normal-type oxygen atoms (see box 10.5 on isotopes), O-18/O-16, where warmer temperatures correspond to higher values of this ratio (see box 9.1). We can, for example, extract oxygen-containing glacial ice and measure this ratio quite precisely; we then assign a time to each measurement by simply counting the number of annual ice layers below the current year's accumulation of snow.

BOX 9.1 O-18 CONCENTRATION AS A PROXY FOR TEMPERATURE

Molecules are constantly in motion. Temperature is, as noted earlier, simply a measure of the energy in this motion. This energy is shared equally among any large collection of molecules. Imagine a billiard table with not only billiard balls but ping pong and bowling balls as well. If you hit the cue ball toward the others and they scatter, the bowling balls will move very slowly, the billiard balls will roll at an intermediate speed, and the ping pong balls will simply fly off the table at high speed. The energy in the cue ball has been shared equally, but the rate at which each ball moves depends on its mass; the heavy things move slowly and the light ones more rapidly.

The same is true in the molecules of air or water—the more massive the molecule, the slower it moves at any given temperature. Thus, the water molecules that contain two hydrogen atoms (one mass unit each) and one O-16 atom move more rapidly than the water that has two H's and an O-18 atom (which has $20/18 = 11$ percent more mass). When water molecules near the surface of the ocean are jockeying to see which will break free and evaporate into the air, the fast-moving molecules have a significant advantage. Thus, the water vapor that ascends to the clouds—and then eventually falls as snow on the glaciers—has an over-representation of O-16 relative to O-18.

Now, to oversimplify the story somewhat, if the Earth's temperature rises, all the molecules in the ocean speed up. This makes it easier for some of the O-18-containing water to also break free and join the clouds . . . and then the ice. The greater the concentration of O-18 in the glacial ice, the warmer the temperature at the time of its evaporation. Thus, the O-18 concentration acts as a thermometer, providing the history of Earth's temperature over hundreds of thousands of years.

The result can be presented as a standard time series plot of the O-18/O-16 ratio versus time. This is a literal representation of the measurements. But the thing of interest is temperature versus time. By carefully comparing the O-18/O-16 ratios from the last 125 years during which simultaneous, direct thermometer measurements of temperature are available, we can "calibrate" our proxy; i.e., we equate various oxygen ratios to their respective temperatures. We can then plot the real data (oxygen ratio vs. time) and the quantity of interest (temperature vs. time) on the same graph, labeling the left vertical axis of the plot with a temperature scale (see figure 9.1).

Proxies are invaluable in our quest for a quantitative understanding of Nature. It is important, however, that we choose proxies carefully and that we never confuse the proxy with the real quantity of interest. A good proxy has a direct connection to the datum we want to know and is not

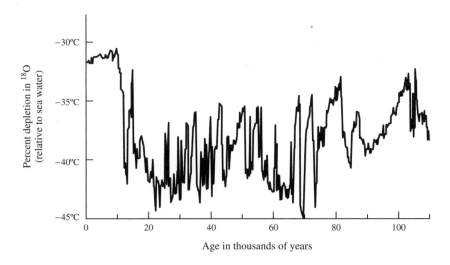

FIGURE 9.1   TEMPERATURE FOR O-18 DEPLETION IN GREENLAND ICE

The depletion of O-18 in the Greenland ice relative to today's sea water over the past 110,000 years plotted as the inferred temperature.

heavily influenced by other, extraneous factors. The latter criterion may be difficult to satisfy, and the confounding influences may remain hidden from view for a long time. For example, for more than a quarter of a century, astronomers used the relationship between the pulsational period of a certain class of variable stars and the stars' energy output to infer their distances. Because these stars are very luminous and can be seen across great distances, they provided the basic meter stick we used to measure the size of the universe. But the proxy of pulsation period for actual luminosity (and thus distance) was corrupted by the fact that observers failed to recognize that stars with different chemical compositions behave differently—chemically distinct stars can exhibit the same period of pulsation despite having very different luminosities. When the error was finally realized, our estimate of the size of the universe doubled.[6]

All proxies involve assumptions (discussed later in this chapter), and the best way to avoid some of the pitfalls that proxies introduce is to be as explicit as possible about what assumptions underlie the substitution of the proxy for the real quantity of interest. Used appropriately, proxies extend the reach of science far beyond the bounds of space and time that we can observe directly in the laboratory.

SELECTION EFFECTS

Ensuring the integrity of data is one of a scientist's foremost responsibilities. Systematic errors and statistical uncertainties of each measurement must be reduced as much as possible. Care must also be taken to ensure that the data obtained represent a fair sampling of the phenomenon under study. This latter requirement does not come naturally to humans. As noted in chapters 5 and 6, our brains have evolved to become highly adept at finding patterns but not at gathering comprehensive and scrupulously unbiased data. We are always ready to find the remarkable coincidence in the fortune cookie's prediction and the events of our day, ignoring all the events to which it had no connection. We note the eerie repetitions of history (Presidents Lincoln and Kennedy were both elected to the House of Representatives in '48, to the presidency in '60, and both were assassinated sitting beside their wives; both had a vice president from the South, born in '08, and named Johnson with a six-letter first name) and we ignore the thousands of incongruities (what about Lincoln's wife, kids, butler, and secretary of the treasury?). In science, such blinkered observations lead to what we call "selection effects," or "selection bias."

Selection effects can come in many guises. The most obvious, and most egregious, is selectively reporting the results of one's experiment or observation, displaying only those data that support one's hypothesis and discarding the rest. If done deliberately, this is considered scientific misconduct; if done inadvertently, it must be labeled incompetence. A more common bias arises in situations where it is only possible to measure part of the whole range of conceivable properties. This can result from the limitations of one's instrumentation. For example, the physical nature of light is such that the ability of a telescope to distinguish two nearby objects depends only on the wavelength of light being observed and the diameter of the telescope mirror; no cleverly designed camera or meticulous observing technique can overcome this physical limitation (see chapter 3 and figure 3.1). Thus, in a survey to see how many nearby stars have companions, the size of the telescope being used limits our ability to see stellar pairs separated by less than our telescope's particular limiting angle. Closer pairs may well exist, but we can't detect them; thus, we will systematically underestimate the number of nearby double stars.

The most subtle types of selection bias arise when some variable beyond our knowledge or control is hiding a subset of measurements from view. In conducting a telephone survey for a polling organization on yacht ownership in Manhattan, two classes of people will be missed completely: those without phones and those with unlisted numbers. Whereas the former group might not be expected to represent a large yacht-owning class, the latter may well represent the majority of such people, and the survey's value would be seriously compromised. When counting lemur species in Madagascar by day, one will inevitably miss the strictly nocturnal species. If one can recognize the possibility of such hidden data, a new experimental approach might eliminate the selection effect; e.g., obtaining the state registry of yachts and culling all those with Manhattan addresses could solve the unlisted phone number problem. But if Nature evinces a phenomenon whose existence we have not yet recognized (such as the dust-hidden quasars discussed in chapter 5), the selection bias may go unrecognized for decades.

Another kind of selection effect arises from the sociology of scientific publishing. Scientists are people, and people like to report positive results. Likewise, journals like to publish positive results. Suppose you decide to perform an experiment using your friends to test the popular (although largely exaggerated) left brain/right brain dichotomy. Each day, for three weeks, you buy a different friend an ice cream cone and then surreptitiously watch how they eat it. Do they sculpt it into elaborate shapes with their tongues, producing veritable ice cream art, or do they just systematically munch through it? You find both behaviors and score each person on a scale of 1 to 10, where 1 is a straightforward muncher and 10 has the tongue technique of Rodin's fingers. You also note whether these friends are right- or left-handed (and thus, supposedly, left- or right-brained). When you compile all your data, it looks as though you've wasted $60 on ice cream. There is no apparent connection between ice cream consumption styles and the putative dominant brain types. To avoid wasting more time, you do not bother to write up the results for publication.

But you did do an experiment: you collected data and got a result. Now imagine that thousands of other curious scientists have done similar experiments on left-brain/right-brain phenomena. And suppose that 99 percent of them got null results just like you and didn't bother to publish. That may still leave several dozen studies that yielded positive correlations, even if there is no real correlation at all—if you do enough trials, a few will look significant (if you flip a coin often enough, you are bound to get ten heads in a row). The problem arises when only the positive results get published. The consensus in the scientific literature would be clear: human behavior is governed by which side of your brain is dominant. But is it?

Although the situation I have described is extreme (and somewhat unrealistic), there is little doubt that publication bias exists and is a significant problem in some fields. Studies that examine many papers about a phenomenon in an attempt to discover small trends that might not be apparent in any single experiment (commonly referred to as "meta-analysis") often take publication bias into account explicitly to avoid reaching erroneous conclusions. Fortunately, in most scientific fields, generalized skepticism is sufficiently strong that surprising results are tested many times, and negative conclusions do get published—theories get falsified. In some fields, however, the situation is extremely serious (see chapter 11), so much so that publication bias contributes to, rather than militates against, the Misinformation Age.

EXPERIMENTS AND OBSERVATIONS

The range of scientific inquiry is vast—from the quarks inside a proton to the edge of the visible universe or the behavior of a single nerve cell to the evolution of an ecosystem. Unsurprisingly, the kinds of scientific tools employed and the modes of data collection vary widely from discipline to discipline. There is one broad dichotomy, however, that is worth a brief comment: experiments versus observations. In many areas of physics, chemistry, molecular and organismal biology, psychology, neuroscience, etc., scientists have their hands (or tools, which are the extension thereof) on their subject matter. In such circumstances it is possible to do experiments: human-created situations in which the environment is controlled, the start and stop time are carefully scheduled, the manipulations are of the experimenter's design, and the results spew forth in a predetermined format. In astronomy, oceanography, climate studies, ecology, and some other scientific endeavors, the systems under study are so remote in space and/or time or are so vast and complex and uncontrollable, the scientist is often reduced to making observations.

Both experiments and observations, if designed well, can yield quantitative and unbiased data concerning the behavior of the natural world. But observers have less control over the environment in which their subject is operating and must take extra care to minimize (or measure) the effects that the variables beyond their control may have on the data of interest. Nineteenth-century naturalists in Britain collected volumes of data on local bird species and meticulously measured their physical attributes, breeding and feeding habits, migratory timings, etc. But with a large heterogeneous population of birds, a strong seasonal variation in climate, multiple food sources, and a broad distribution of natural topography and human land use, the factors that determine things such as beak size were far from apparent. In the Galápagos, Darwin found a very limited population of finches, isolated from their parent population for a long time, living on a few small islands with different microclimates and vegetation. The striking differences in finch species found in different locations formed a cornerstone of his subsequent theory of evolution through natural selection. For observers, it is a constant battle not only to identify all the influences affecting the object of study but also to discern which are relevant for the question under consideration and which can be safely ignored.

MODELS

Models provide the conceptual framework for interpreting the data we collect. They come in four different forms:

► Physical: A rotating tank with two fluids of different densities can be used to study the onset of turbulence that leads to magnetic storms on the Sun.

► Biological: A mouse engineered to have the CAG repeats on gene ITIS, the coding error that produces Huntington's disease, can be used as a model for humans with this condition (biologists call these models knock-in mice).

► Numerical: The millions of lines of computer code that make up a general circulation model capture the complex interaction of oceans, ice sheets, atmosphere, and plants to predict the future course of Earth's climate.

► Analytical: Newton's simple equation describes the flight of a ball after it leaves a bat and allows us to calculate where it will land.

Models explain data in the very limited sense that they provide a way (note that I say *a* way, not *the* way) of systematizing and describing what can be a very large number of measurements in a relatively compact form. In addition, if they're good, models can predict how a natural system will respond if looked at, tweaked, or probed in a different way: suppose we look at the fly ball from the stands instead of from behind home plate, smear the ball with a little grease before throwing it, or play the game on the Moon instead of in the Bronx. A good model will still get the outcome right—it is either a home run or it is not. The datum shows that the ball either lands in the stands or on the field. The model is good if it can predict the outcome in advance.

It is always important to keep in mind that a model is not the real thing—a model airplane is not an airplane (it can't get you from New York City to San Francisco in six hours), and the model of a physical situation or event is not the same thing as that situation or event itself. The model is an abstract human construction that attempts to incorporate the essential ingredients of a natural process or system and then to make predictions about future behavior. We strive to include all the important parameters and the relationships among them but, inevitably, we use approximations to do so and incorporate assumptions along the way. As a result, the model predictions have an associated uncertainty. Just as it is critical to assign realistic uncertainties to our measurements, it is essential to recognize and to quantify the uncertainties in our model predictions. Only then can a statistically meaningful comparison of the real world to our model world be made.

THEORIES

When a model has developed to the point where it has successfully predicted the outcome of many different kinds of experiments, has made lots of testable predictions, and has been around for some time, it can gradually morph into a theory (although since there are no clear boundaries that separate models from theories, whether or not something is called a theory is sometimes related to the size of its proponents' egos). The theory of evolution allows us to understand the transformation of slime molds into scientists: start with RNA, add energy, allow mutation, follow natural selection, and voilà—a human being. There are lots of details, of course, that this theory does not specify, but it makes lots of predictions that tens of thousands of evolutionary biologists are testing every day, and, remarkably, it seems to work.

When a theory is around for a long time and continues to explain successfully a large body of data, it is sometimes promoted to the status of a law of Nature. I have never much liked this formulation myself. Humans invented the notion of "laws" to regulate social behavior, and the "rule of law" is often cited as an important element in the advancement of human civilization. However, Nature, the physical universe in which we live, has neither presented us with an inventory of its laws nor is it bound to follow the ones scientists invent. Newton's universal law of gravitation was spectacularly successful for nearly three centuries at explaining ever-more refined measurements of the motions of the moons and planets as well as describing motion on the Earth's surface; indeed, the term "universal" was meant to tie together earthly and celestial phenomena with a single theory, a radical break with Aristotelian philosophy, which regarded the two as fundamentally separate. But this law, as it turns out, is neither universal nor right. Careful measurements of the orbit of Mercury show that this planet does not follow the law's dictates. Einstein's general theory of relativity, completely different both in its conceptualization of gravity and the mathematics that describes it, has now displaced Newton's law as our theory of gravity. Perhaps out of appropriate humility, the general theory remains just that—a theory, not a law.

This is not to imply that Newton's law is useless. Indeed, when we need to send a space probe a billion miles through space to explore Saturn, we calculate the trajectory using Newton's law of gravity. Newton's law is a very good approximation to the way Nature works—meaning it is a good theory. General relativity is a better approximation because it subsumes all that Newton's theory predicts and does a better job of describing Nature where gravity is strong (such as near the Sun in the case of Mercury's orbit and under the more extreme conditions near such bizarre celestial objects as neutron stars and black holes). But we still regard general relativity as an approximation—it does not, for example, allow us to say what must have happened in the first 10^{-35} seconds after the Big Bang, implying that it too is an incomplete description of Nature. Nonetheless, it is still a very good theory.

It is important to note that for a scientist the word *theory* has a precise meaning: a usually mathematical formulation describing testable and predictive models of the world. It is neither a synonym for speculation, such as "I have a theory about why the Red Sox always end up losing to the Yankees", a nebulous concept, such as "I heard this theory that rocket launches affect the weather"; nor a cover for reluctance, such as "Well, in theory I would like to go with you to the all-day Gregorian chant festival, but I have to count the number of sheets of toilet paper we have left to see if I need to buy some." The epithet frequently hurled by creationists, that evolution is just a theory, completely misses the point. Evolution is indeed a theory—and, as such, lies at the heart of a scientific approach to the world.

ASSUMPTIONS

It is difficult to get up in the morning if you do not make assumptions. You assume the floor will be there when you roll out of bed and that the air you are comfortably breathing while supine exists several feet higher in the room as well. You assume that the bathroom will be where it was last night and that when you mix hot water with dark brown grains you'll experience the familiar sensation of a cup of coffee. When you get in the elevator, you assume it will descend in a controlled way and deposit you gently in the lobby. These assumptions are all based on experience; they will be borne out by your experience and will thus reinforce your expectation that you can make the same assumptions again tomorrow. Indeed, they have by now receded into your unconscious; you don't recognize them as assumptions at all.

In science, a vigilant awareness of one's assumptions is essential. This is not to say that scientists do not carry around the same unconscious assumptions about elevators and bathrooms as you do, but in gathering data, constructing a model, or attempting to falsify a theory, rigorous attention to all relevant assumptions is required. Phrenology—the inference of personality traits from the study of bumps on the skull—started from the assumption that particular parts of the brain controlled particular behaviors. This assumption has been borne out by a century of work in neuroscience. However, phrenology also assumed that the size of each brain region determined how much of a particular behavior a person would express and, furthermore, that the sizes of different regions were reflected by protrusions on the skull. Both of these assumptions are wrong, explaining the demise of phrenology as a key to human behavior.

Most assumptions are testable, and many are quantifiable. Early global climate models assumed the Sun's energy output was constant. This assumption was consistent with the available measurements taken from the Earth's surface, but such measurements are only accurate to a few percent. Now that we have accumulated thirty years of satellite data on the Sun, we can see fluctuations in its output at the level of two-tenths of 1 percent over its eleven-year sunspot cycle—and we can relax our assumption of constancy and incorporate the changes in the model. However, if our goal is to reconstruct the dramatic changes in Earth's climate over the past million years, we must revert to our assumption of long-term constancy and recognize the limitations this assumption imposes on our results.

Some assumptions in science last a very long time. From Aristotle to the physicists at the turn of the twentieth century, it was assumed that space was filled with aether. To the Aristotelians, this was a philosophical preference, since they assumed that Nature abhors a vacuum. To James Clerk Maxwell, who unified electricity and magnetism into our modern theory of light, the aether was a necessary assumption—the light waves needed a medium through which to travel. But when Albert Michelson set out to measure the properties of the aether, he discovered it was not there—a two-thousand-year-old assumption swept away, paving the way for the relativity revolution.

Scientists question everything—especially their assumptions.

FEEDBACK EFFECTS

It is a standard feature of school assemblies: the principal steps to the microphone, begins to speak, and an ear-piercing shriek emerges from the sound system. Feedback. What's actually going on?

The sound system includes three principal components: a microphone, an amplifier, and a speaker. The microphone picks up the tiny vibrations of air molecules that constitute the sound of a voice and transforms (transduces is the technical word) them into an electrical current. The amplifier magnifies the size of this variable current and passes it along to the speakers, where the electricity is transduced back to sound by driving membranes to vibrate back and forth, jiggling the air molecules in synchrony with the sounds being picked up by the microphone. If the microphone receives some of these increased vibrations from the speakers, it obviously records them as well and passes them on through the amplifier to be boosted again, raising the sound volume coming out of the speaker and leading to greater microphone input. This feedback loop of ever-increasing sound soon saturates the amplifier, leading to the predictable shriek.

Many natural (and social) phenomena exhibit feedback, which can be described succinctly as a process wherein the effect affects the cause; that is, the outcome of some physical process, some biological activity, or some social interaction either enhances or suppresses that process, activity, or interaction. In the case of the microphone, the amplification and reproduction of the sound originating from the input to the microphone results in a greater amplitude of sound at that input. This is called "positive feedback": the effect amplifies the cause. Despite its name, positive feedback often leads to negative consequences. Consider the following feedback loops:

Unemployment rises → income tax revenues fall → government expenditures for construction projects fall → unemployment rises further

The Earth's temperature rises → more water evaporates into the atmosphere → more solar radiation is trapped near the Earth's surface → the temperature rises further

Your little brother hits you → you are irritated → you hit him back → he gets mad → and hits you harder

Alternatively, the effect can suppress the cause; this we call "negative feedback." Note that this usage is not the same as the colloquial sense of this term: "My teacher gave me negative feedback on my paper; it was covered with red ink." Rather, negative feedback is a technical description of a process in which the result damps down whatever triggered the process in the first place. For example:

Unemployment rises → disposable incomes fall → lower demand depresses prices and lowers wages → exports are cheaper → demand for exported products rise → people are hired to make the exported goods → unemployment falls

The Earth's temperature rises → more water evaporates into the atmosphere → more clouds form → more sunlight is reflected back to space → less solar energy penetrates the atmosphere → the Earth's temperature falls

Your little brother hits you → you feel a pang of guilt for all the frustration your success has bred in him → you hug him and slip him a $20 → he smiles and doesn't hit you again all day

Negative feedback often has positive consequences. Whatever the value judgment associated with the outcome, however, negative feedback is a stabilizing influence, whereas positive feedback tends to lead to runaway situations.

The natural world, along with the worlds of geopolitics, economics, and social intercourse, are rife with positive and negative feedback loops, often running simultaneously. In building models of these phenomena, we must be cognizant of feedback and take care to incorporate it appropriately. One of the more extreme examples of a system dominated by complex, interacting feedback loops is the global climate system. The fidelity of current global climate models is limited not so much by the accuracy of the equations employed or enough data to constrain them as it is by the many subtle feedback effects that must be considered in how the Earth's ice, oceans, land, air, and biology interact in the presence of a steady flow of energy input from the Sun. Understanding feedback is essential for progress in modeling almost all complex natural (and social) systems.

In fact, the issue of global climate changes provides an excellent case study in which to apply the habits of mind that characterize science. It is also a striking example of how much misinformation pervades our age and why a survival guide might be needed. In the following chapter, I provide an outline of some of the issues pertinent to the subject of climate change, illustrating as I go the features of the scientific approach I have sought to convey in the preceding pages.

10
APPLYING SCIENTIFIC HABITS OF MIND TO EARTH'S FUTURE

n an interview with the *New York Times* discussing the possibility of his seeking the 2016 Republican nomination for the presidency, Florida's junior Senator Marco Rubio was asked about climate change. He responded by saying he disagreed with scientists that humans were responsible for what he referred to as the "always evolving" climate: "I do not believe human activity is causing these dramatic changes to our climate the way scientists are portraying it," he said. "And I do not believe that the laws that they propose we pass will do anything about it, except destroy our economy."[1]

Two days later, another one of the Senate's "experts" on the Earth's climate, Senator James Inhofe of Oklahoma, author of *The Greatest Hoax: How the Global Warming Conspiracy Threatens Your Future*, weighed in. To establish his credentials, let me first quote from an interview promoting his book on the Voice of Christian Youth America's radio program *Crosstalk*, where Inhofe stated:

The Genesis 8:22 that I use says that "as long as the earth remains there will be seed time and harvest, cold and heat, winter and summer, day and night." My point is, God's still up there. The arrogance of people to think that we, human beings, would be able to change what He is doing in the climate is to me outrageous.[2]

In response to the National Climate Assessment 2014 report, coauthored by Admiral David Titley, a meteorologist who now directs a climate center at Penn State, Inhofe stated:

There is no one in more pursuit of publicity than a retired military officer. I look back wistfully at the days of the Cold War. Now you have people who are mentally imbalanced, with the ability to deploy a nuclear weapon. For anyone to say that any type of global warming is anywhere close to the threat that we have with crazy people running around with nuclear weapons, it shows how desperate they are to get the public to buy this.

Titley's response: "The ice doesn't care about politics or who's caucusing with whom, or Democrats or Republicans."[3]

In 2012, the North Carolina state legislature passed a law that bans the state from using scientific predictions about sea level rise to develop its coastal policies. A major proponent of the law was Tom Thompson, president of NC-20, a group of coastal developers who denies that global warming exists. During an episode of *The Colbert Report* shortly thereafter, comedian and host Stephen Colbert satirized the law quite appropriately: "If your science gives you a result you don't like, pass a law saying the result is illegal. Problem solved."[4]

And so it goes.

Like Senators Rubio and Inhofe, I am not a climate scientist. I have no grant funding to study climate change and no tenure committee vetting my work to decide on my future employment. As a scientist, however, I do have broad interests. Several decades ago while preparing a new course for Columbia undergraduates, I read a number of journal articles on paleoclimatology, the use of various **proxies** to reconstruct the history of the Earth's climate, and became fascinated by the subject. Over the past quarter century, I have kept up with developments in the fields of both paleoclimate and climate change. A few years ago, as part of a series of public lectures I was to give, I spent several weeks preparing one entitled "Global Warming: What We Know and What We Don't Know." My goal was to approach the problem as an appropriately **skeptical** scientist, using my knowledge of physics and astronomy where appropriate and reading a broad range of papers on aspects of the problem about which I was less familiar.

Here is the result. Throughout this chapter, I highlight in **bold** the words and concepts that I have been presenting as characterizing scientific habits of mind. The result, I hope, is an illustration of how to approach information and misinformation and how to distinguish between the two; as a bonus, a rational presentation of the climate change issue may be of general interest.

SCENE 1

I open my lecture by showing trailers from two films: *The Day After Tomorrow*, a Hollywood disaster movie in which the Earth's climate system goes haywire in the course of a week, wreaking havoc across the globe; and Al Gore's *An Inconvenient Truth*. What I find noteworthy about these two trailers is not the differences but the similarities. Lots of jump cuts, dramatic music, poignant human tragedy, sound-bite length statements of "fact"—in short, a blatant appeal to primitive instincts and emotions.

My first slide (figure 10.1) is meant to invoke a contrasting approach. I explain how each subsequent slide will be identified with a label that belongs to one of the following five categories:

> Facts: For these purposes, to avoid epistemic arguments, I define facts as **measurements** of some physical quantity that are conducted with the best available tools and vetted by **skeptical** review.

FIGURE 10.1 INTRODUCING GLOBAL WARMING

The title slide of my presentation on global warming.

Physics: I emphasize that physics provides **models** for many of the relevant phenomena—indeed, the best **models** that humans have yet invented—but should never be confused with physical reality itself.

Feedback: As noted in chapter 9, **feedback** plays a crucial role (and constitutes one of the largest uncertainties) in building climate **models**.

Foreshadowing: Foreshadowing hints not only at the future but also illustrates where our **models** fall short.

Fiction: For this category I provide a number of examples from the public debate (although far from an exhaustive catalog).

SCENE 2

I begin by asking the obvious question: What determines the temperature of a planet? If we want to understand whether it is going up or down, it would help to have an understanding of what sets a planet's temperature in the first place. As noted earlier, temperature is just a direct measure of the **mean** kinetic energy (energy of motion) of the atoms and molecules that make up a substance. As a consequence, the physics equation that governs the temperature of a planet is simplicity itself: energy in = energy out (figure 10.2).

As you have no doubt learned in school, the Sun is the primary source of the Earth's energy, completely dominating all other energy inputs

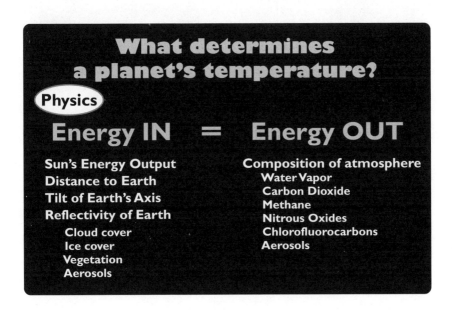

FIGURE 10.2 TEMPERATURE FROM THE ENERGY BALANCE EQUATION

A representation of the basic energy balance equation that determines a planet's temperature and the factors that control it.

(see box 10.1). So the Sun dominates the energy-in side of the equation. What you might not have realized is that the Earth is radiating into space an amount of energy equal to that which it receives. If it didn't, it would be like an oven without a thermostat, and the Earth would just get hotter and hotter until it evaporated. The temperature of any planet is set by this balance between incoming and outgoing energy; any imbalance changes the temperature.

The parameters we must consider that affect the energy received include the Sun's energy output (clearly if this goes up or down the energy we receive will go up or down), the distance between the Sun and the

---

BOX 10.1 ENERGY SUPPLY AT THE SURFACE OF THE EARTH

The **average amount** of energy received by each square meter of Earth from the Sun (taking into account day and night, variations in latitude, etc.) is about 342 watts, where one watt is the energy of one joule (the metric unit for energy) received each second. A 100-watt light bulb uses 100 joules of energy per second, so the Sun's energy supply to Earth is the equivalent of about 3.5 100-watt light bulbs for each square meter (about 11 ft.^{2}). Integrating over the Earth's surface (taking its **mean** radius to be 6,371 km) yields 174,000 terawatts, where a terawatt is 10^{12} (one **trillion**) watts.

The other sources of energy that impinge on Earth's surface are both constant and tiny compared with the Sun's input. They include moonlight and starlight, the remnant heat from the Earth's formation that seeps up from below, radioactive decay of nuclei in the Earth's crust, and junk falling in from space. If we sum all the nonsolar contributions, we get a total of a little under fifty terawatts or less than 0.03 percent (0.0003) of the energy received from the Sun. Furthermore, since these contributions don't change appreciably with time, it is safe to ignore them in discussing climate change.

Earth (the **farther** away we are, the smaller a fraction of the Sun's energy we receive), the tilt of the Earth's axis (which actually affects the distribution of energy over the surface rather than the total energy but also has consequences because of **feedback loops**), and the reflectivity of the Earth (light reflected directly back into space makes no contribution to heating the Earth). The reflectivity term depends on such factors as cloud cover, ice and snow cover, vegetation, and dust particles in the air.

I address each of these factors in turn. For illustrative purposes, I will review just one of them here—the solar output. The next slide (figure 10.3) presents a **fact**: the Sun's total energy output is 3.839×10^{26} watts.

The question to ask is whether it changes. Figure 10.4 provides the answer in the form of two **time series plots**. The best **data** are from NASA

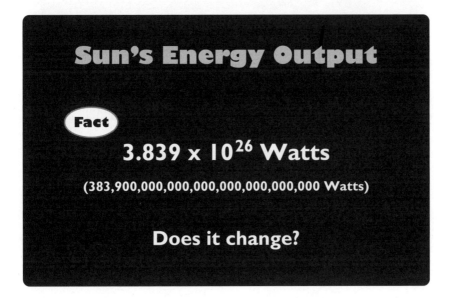

FIGURE 10.3 THE SUN'S ENERGY OUTPUT

The **mean** energy output of the Sun in metric units of Joules/second = watts.

satellites that have monitored the Sun's output from above the obscuring effects of the Earth's atmosphere for decades. The **plot** shows both day-to-day variations of a few tenths of one percent, and a regular, eleven-year pattern of lower **mean** energy released (and smaller fluctuations) followed by a **mean** about 0.1 percent higher with greater short-term fluctuations. These changes are a consequence of sunspots, first discovered by Galileo in 1610 and now understood to be magnetic storms on the surface of the Sun that wax and wane in intensity and frequency with an eleven-year cycle. Although sunspots appear dark to the eye (implying less energy release), they in fact generate more total energy in the ultraviolet and x-ray portions of the spectrum, so the net result is an increase in solar output, as shown in figure 10.4.

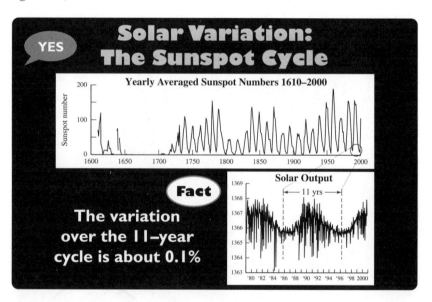

FIGURE 10.4 SOLAR OUTPUT VARIABILITY

Measures of solar output variability. The bottom plot shows actual measurements of energy received from satellite **data**, whereas the top curve shows the history of sunspot counts since their discovery in 1610.

Also shown in figure 10.4 is a much longer record that may provide a **proxy** for solar variability—the number of sunspots recorded over the last 400 years. Here we see variability in the maximum number of spots during an eleven-year solar cycle of not tenths of a percent but a factor of ten (even if we ignore the seventeenth century when the spots seem to have disappeared entirely). As with all **proxies**, sunspot number is less than ideal. We have no information on the size or lifetime of the spots (both of which **correlate** with the amount of excess energy they produce). And since the thirty-year record we have that overlaps with the satellite observations shows a rather stable number of spots at both the maximum and minimum of the cycle, we don't know if the larger fluctuations seen earlier in the record really correspond to bigger changes in solar output.

We do have other information to complement the raw sunspot number **data** that is suggestive, however. The apparent disappearance of spots in the seventeenth and early eighteenth century, known as the Maunder minimum, corresponds to the coldest period in the last 2,000 years in Europe (see figure 10.10), and the minimum in the eleven-year running **average** seen in figure 10.4 around 1810 is proximate to the unique event of 1816 known as "the year without a summer," which saw snow in June in New York and produced a food crisis in the United States, Canada, and Northern Europe. A major contributor to this event was the explosion of the Tambora volcano in Indonesia the year before that blanketed the Earth in a dust cloud, significantly reducing the solar energy absorbed, but the coincidence with this minimum in the sunspot number (called the Dalton minimum) may also be important.

In summary then, we know with great **accuracy** (and therefore quote with a **precision** of four **significant figures**) the **average** solar energy output. We also know this output varies (a) on short timescales (days to weeks) by several tenths of a percent, (b) in a systematic way over eleven years by a **mean** difference of about 0.1 percent, and (c) possibly on longer timescales (decades to centuries) by a greater amount. To calculate the effect of these variations on Earth's temperature will require taking into account several other effects, **feedback** loops among them (particularly those that affect the Earth's reflectivity—see box 10.2), and building a **model** that takes all of this into account.

BOX 10.2 REFLECTED LIGHT

Rather than being absorbed by molecules in the atmosphere, in oceans, or on land, incoming solar energy reflected directly back into space is energy lost for warming the Earth. As planets go, Earth is quite shiny, reflecting a full 31 percent of the incoming light (54,000 terawatts). This fraction depends on the fraction of the Earth covered with clouds (which varies by the hour), the fraction of land covered with snow and the sea covered with ice (which varies a lot depending on the month of the year), the fraction of land covered with plants (which have evolved to be efficient absorbers of solar energy, unlike deserts, which reflect a lot of light—this also varies with the seasons), and the amount of dust (from dust storms, volcanic eruptions, coal-fired power plants, and automobiles) in the atmosphere. This latter component, grouped together as "aerosols," is particularly **uncertain**, variable, and complicated because molecules such as sulfuric acid (produced in the stratosphere by erupting volcanoes) are highly reflective and produce cooling, whereas black soot particles actually absorb light and thus warm the atmosphere.

Many of these factors are embedded in complex **feedback** loops with the temperature. An example of **positive feedback** is when the Arctic sea ice melts because of warmer temperatures. This results in more exposed ocean (which absorbs sunlight) and less ice (which reflects light efficiently), meaning more energy is absorbed, the temperature rises further, and more ice melts. Alternatively, **negative feedback** occurs when higher temperatures lead to greater ocean evaporation and thus more clouds that reflect sunlight back into space and lower the temperature. More water vapor in the air and higher temperatures may make plants

---

(continued)

in some regions thrive, increasing absorption and raising temperatures further, whereas abundant plant life in marginal zones such as the African Sahel (the arid region just south of the Sahara) can lead to more people and goats that overgraze the plant life, returning the land to a desert and leading to a population crash (as happened in the 1970s and 80s).

The conclusion: **feedback effects** make things complicated. Leaving them out of a **model** can lead to spurious results.

In my presentation, I go through each of the other effects on the energy-in side of the equation, noting the facts, any variations in the measured quantities, **feedback** effects, and where we are relying on physics and **models** rather than **data**. It turns out that we have an extremely accurate **model**, well-**calibrated** with centuries of **data**, that predicts the various orbital effects that modulate the amount of energy the Earth actually receives. This leads to the discovery that there are three quasi-regular cycles of variability to solar input on time-scales of 23,000 years (known as precession), 41,000 years (known as obliquity change—an increase and decrease of the tilt of the Earth's axis), and just over 100,000 years (in the changing ellipticity of the Earth's orbit). These are summarized in the three **time series plots** shown in figure 10.5. I note here that all three effects are going on simultaneously, all three effects are completely beyond our control (we are neither now nor are we in the conceivable future likely to be capable of removing Jupiter from the solar system because it messes with the Earth's orbit shape), and that all three have a profound effect on the Earth's climate, a conclusion confirmed by the **data** I present in the next section.

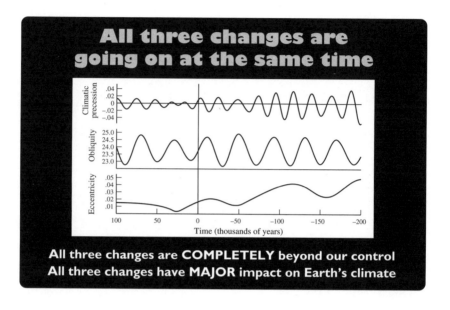

FIGURE 10.5 ORBITAL PERTURBATIONS FOR THE EARTH

Variations in the Earth's orientation and orbital motion over the past 200,000 years and the next 100,000 years.

SCENE 3

The energy-out side of the equation is completely dominated by one factor: the composition of the Earth's atmosphere. Here again the raw data—the facts—are pretty easy to come by. We can easily take a sample of air and, by sorting and counting the various atoms and molecules present, determine the atmosphere's chemical composition. Figure 10.6 shows the data for the dozen most common and important components. Five of these, indicated in gray, are the so-called greenhouse gases, and their presence helps regulate the temperature of the Earth.

Because light that comes from the Sun arises from atoms at a temperature of 5780 K, it is concentrated in the visible part of the spectrum

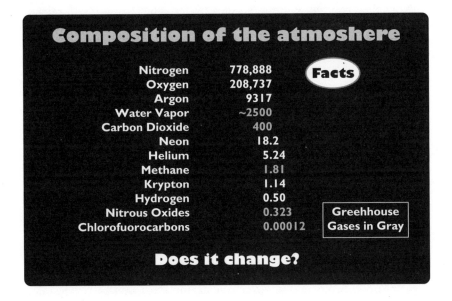

Composition of the atmoshere

Nitrogen	778,888	**Facts**
Oxygen	208,737	
Argon	9317	
Water Vapor	~2500	
Carbon Dioxide	400	
Neon	18.2	
Helium	5.24	
Methane	1.81	
Krypton	1.14	
Hydrogen	0.50	
Nitrous Oxides	0.323	Greehhouse
Chlorofuorocarbons	0.00012	Gases in Gray

Does it change?

FIGURE 10.6 COMPOSITION OF EARTH'S ATMOSPHERE

The concentration of a dozen significant gasses making up the Earth's atmosphere. Greenhouse gasses are shown in gray.

(see chapter 3). As a consequence of our atmosphere's composition, it is (absent clouds) almost completely transparent to visible light, so the sunlight streams through and is absorbed by the ground, water, plants, and beachgoers foolish enough to want a tan (and skin cancer later). All this absorbed energy goes into making all these molecules jiggle faster—that is, it raises the temperature. By virtue of their jiggling, the molecules can also radiate energy characteristic of their temperature, which, **averaged** over the Earth and the seasons, is more like 15°C (59°F) or 288 K, twenty times cooler than the sun. This means that outgoing radiation from the Earth emerges at a wavelength twenty times longer in the infrared (IR) portion of the spectrum.

Now, the greenhouse gas molecules (water vapor being the dominant contributor) are transparent to the incoming visible light but are opaque to the outgoing IR radiation. They thus absorb the energy trying to leave, trapping it in the atmosphere and raising the temperature of the air (and, in turn, other components of the Earth's surface). When the temperature is high enough that the emitted wavelengths are short enough to slip through the greenhouse gas absorbers, the energy can leave. Thus, these gases act as an electric blanket—a warming layer with a thermostat—regulating the global temperature. And it's a good thing they're present; if they weren't, the Earth's **mean** surface temperature would be about $-15°C$ ($5°F$), too cold for life as we know it to evolve. As with any thermostat, however, if you turn it up too much, you get hot.

As with the energy-in side of the equation, we must ask whether the atmospheric composition changes. And as with the other terms, the answer is yes. A particularly dramatic example is given by the Keeling curve, the **time series plot** of monthly **measurements** of carbon dioxide that was initiated by Charles Keeling in 1958 and has continued to this day by his successors (including his son). Figure 10.7 shows the first three years of his **data**; an annual variation of a little more than 2 percent is apparent.[5] CO_2 concentration peaks in May and bottoms out in October. This is a consequence of the plant life on Earth breathing (see box 10.3).

As a moment's thought will make clear, this annual variation in CO_2 concentration cannot have a significant effect on the Earth's temperature because as the CO_2 falls, the temperature rises (to late summer), and as the CO_2 rises, the temperature falls; in addition, the opposite temperature change occurs in the Southern Hemisphere. Furthermore, this kind of a repeating annual cycle shouldn't produce a systematic, long-term trend in temperature. Looking at the full Keeling curve, however, reveals another change. In figure 10.8, the full fifty-seven years of **data** are presented. The apparently large annual cycle is now dwarfed by a **monotonic** rise over the entire period from a **mean** value of 315 ppm in 1958 to just over 400 ppm

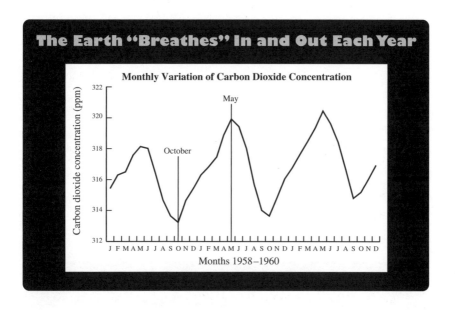

The Earth "Breathes" In and Out Each Year

Monthly Variation of Carbon Dioxide Concentration

FIGURE 10.7 ANNUAL CARBON DIOXIDE CONCENTRATION VARIABILITY

Monthly variation in atmospheric carbon dioxide concentration as measured between 1958 and 1960 by Charles Keeling at the Mauna Loa Observatory in Hawaii.

BOX 10.3 SEASONAL CHANGES IN CARBON DIOXIDE CONCENTRATION

Plants grow by absorbing carbon dioxide from the air, using the energy from the Sun to break the molecule's bond, breathing out the oxygen (which we breathe in), and using the carbon to make cellulose and other plant fibers. In May, the deciduous trees produce leaves, the grass turns green, and crops are planted all over the Northern Hemisphere. These plants beaver away for the next six months, breathing in carbon dioxide to

(continued)

make cellulose, and, consequently, reducing its atmospheric concentration. Then the leaves fall off the trees, the corn stalks are left to brown in the fields, and, over the next six months, bacteria get to work producing decay in all this organic matter by combining the carbon compounds with oxygen from the air. Thus, the carbon dioxide level continues to rise until it is spring again.[6]

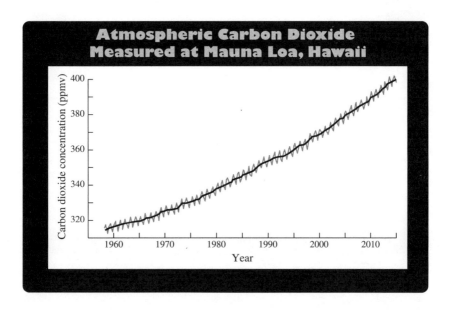

FIGURE 10.8 ATMOSPHERIC CARBON DIOXIDE CONCENTRATION OVER THE LAST HALF CENTURY

The full Keeling curve showing the steady and accelerating increase in carbon dioxide in the Earth's atmosphere.

BOX 10.4 CALCULATING THE MASS OF CARBON DIOXIDE INCREASES

The entire Earth's atmosphere has a mass of about 5×10^{18} kg (which can be derived if you know the Earth's radius [6,371 km], the height below which about half the atmosphere resides [~5 km], and the density of air, which is about one-thousandth the density of water—can you do it?). From the curve, we can see that between 2003 and 2013, the CO_2 concentration increased by about 25 ppm, so $25 \times 10^{-6} \times 5 \times 10^{18}$ kg $\times 1$ ton/10^3 kg = 1.25×10^{11} tons or 125 gigatons (Gt). In fact, human activity now produces about 30 Gt per year or 300 Gt per decade. Thus, it seems that about 40 to 45 percent of the CO_2 we produce ends up in the atmosphere; the remainder is dissolved in the ocean or absorbed by plants.

today, a 27 percent increase. Furthermore, the slope of the line (representing the change per year) continuously steepens.

We can do a **back-of-the-envelope calculation** as to how much carbon dioxide has been added in, say, the last decade (see box 10.4). The answer is roughly 125 gigatons (125 **billion** tons) or roughly 45 percent of the CO_2 that arises from human activity. This demonstrates that we are producing more than enough CO_2 each year to account for the increasing concentration in the atmosphere. It does not, however, prove that our CO_2 production is causing this rise.

To establish our direct culpability, we need a way of identifying the extra CO_2 molecules in the atmosphere. Fortunately nature provides one: isotopes (see box 10.5). It would also be very helpful to have records that extend further back than the fifty-seven years of Keeling's CO_2 **data** or the ~150 years of temperature records we have from systematic thermometer readings if we want to examine long-term climate trends. Again, isotopes to the rescue!

BOX 10.5 EXPLAINING ISOTOPES

Each of the ninety-two types of atoms—called elements—that occur in nature has a unique number of positively charged protons and an equal number of negatively charged electrons; every possibility from one (hydrogen) to ninety-two (uranium) is represented. The specific pattern of electrons for each type of atom determines how it will interact with every other type of atom. Carbon, with six electrons, is most comfortable combining with two oxygen atoms (with eight electrons each)—thus, CO_2 is a very stable molecule. All of chemistry, then, is about the arrangements of electrons.

There is a third, neutral particle in nearly every atom called a neutron. Although neutrons help hold the nucleus together, they have no electrical charge and thus no effect on the chemical combinations atoms can make. And unlike the number of protons and electrons that are the defining characteristic of each kind of atom, the number of neutrons each nucleus has can vary. For example, carbon comes in three varieties: C-12, which has six protons (by definition all carbon atoms have six protons) and six neutrons; C-13, which has six protons and seven neutrons; and C-14, which has six protons and eight neutrons (where the labeling number simply represents the sum of the protons and neutrons in the nucleus). Carbon-12, which makes up 98.9 percent of natural carbon, is most common, whereas C-13 makes up almost all of the other 1.1 percent. C-14 is only present in about one part in a trillion (an abundance of 10^{-12}).

Likewise, hydrogen has three naturally occurring isotopes: H-1, H-2 (called deuterium), and H-3 (called tritium—only for hydrogen do isotopes get individual names). Oxygen has even more isotopes, although the two most common are O-16 (the dominant species) and O-18 (which we will meet again later in this chapter).

The property that distinguishes different isotopes of the same element is their mass—neutrons are even slightly heavier than protons so contribute significantly to the mass of the atom. This difference in mass introduces a subtle form of discrimination into all chemical reactions.

Recall the analogy in chapter 9 of the billiard table with the bowling, billiard, and ping pong balls. When atoms and molecules are running around freely as they are in the atmosphere, the light ones move fast and the heavy ones move slowly. Imagine an oak tree breathing in CO_2 through the little openings on the back of each leaf and ready to use its photosynthetic factory to cleave off the carbon and turn it into wood molecules. The first step in the process is to produce a chain of three carbon atoms linked together. The CO_2 molecules with C-12 atoms are going to be moving just a little faster (by about 1 percent) and so, on **average**, will get through the machinery just a little more readily and thus will be more likely to end up in the wood. The CO_2 molecules that contain C-13 atoms get discriminated against. It turns out that for deciduous trees, the discrepancy is about 2.5 percent; that is, a piece of wood from a tree growing in a temperate climate will have 2.5 percent fewer C-13 atoms in its wood than were present in the air in which it breathed. C-14, heavier still, is discriminated against even more severely.

Fossil fuels—coal, oil, and natural gas—are decomposed, compressed, and transformed plant matter that were laid down in deposits over almost 200 million years during the age of the dinosaurs. The same laws of physics applied back then—C-13 was discriminated against in making plant fibers. Thus, the CO_2 formed from burning fossil fuels (and burning is just combining the carbon in the fuels with oxygen from the air) has less C-13 than the CO_2 in the ambient air.

Trees from temperate climates come with a wonderful natural clock built in—annual growth rings (figure 10.9). The outermost ring represents the current year's growth, with each ring interior counting off the years since the tree was born. By using the oldest living trees on Earth— the bristlecone pines from the mountains in the American West that are

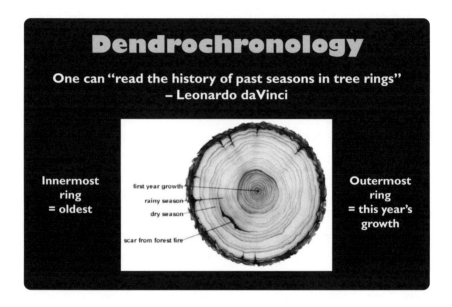

FIGURE 10.9 ANNUAL GROWTH RINGS OF A TREE

The cross-section of a tree revealing the annual growth rings that allow us to read the climate history.

up to 4,500 years old—and combining them with trees used in prehistoric construction projects and those preserved in mud at the bottom of rivers, we have constructed a continuous record that covers every year between the present and 11,000 years ago. Before writing, before numbers, and long before science, trees have preserved a continuous record of the amount of C-13 in the atmosphere. All we need to do is extract a little piece of wood from a ring whose age is known precisely and count the carbon isotopes.

We find a record of constant C-13 over thousands of years until around 1850, when the fraction of C-13 atoms begins to drop. It falls continuously and at an increasing rate to the present day, a time with by far the lowest

C-13 concentration in the atmosphere of the last 11,000 years. The middle of the nineteenth century was when the Industrial Revolution accelerated with massive burning of coal and, a decade later, the discovery of oil. All of this fossil fuel was deficient in C-13 and, as the fraction of the CO_2 in the atmosphere from fossil fuel burning increased, the C-13 isotope ratio steadily declined. C-14 declines even faster than it would have if modern plants were somehow implicated (e.g., burning rainforests) because this isotope is completely absent in fossil fuels, having decayed away over the hundred million years they have been under ground. We see exactly the same pattern in coral reefs that are built up over thousands of years. The fingerprint is crystal clear. The CO_2 accumulating in the atmosphere today comes directly from fossil fuel combustion.

SCENE 4

The extensive tree ring record provides another enormous benefit for the study of past climate—its isotopic signatures can yield an annual record of temperature and humidity as well. As with CO_2, H_2O molecules that contain heavy hydrogen (H-2) or heavy oxygen (O-18) move more slowly than the more common H_2O with H-1 and O-16. As noted earlier, when a water molecule evaporates (goes from a liquid to a gaseous state), it must achieve sufficient speed to break free from its fellow liquid water molecules. In cool years, H_2O-18 is heavily discriminated against because it has trouble reaching the speed necessary to escape. When it is warmer, all the water molecules move faster, and a larger proportion of H_2O-18 can break free. Thus, the O-18/O-16 isotope ratio provides a simple thermometer that records the **average** temperature of that growing season. Through a different mechanism, the C-13/C-12 ratio provides a measure of **average** humidity. Even more simply, measures of ring width and wood density in each ring provide a well-**calibrated proxy** for temperature.

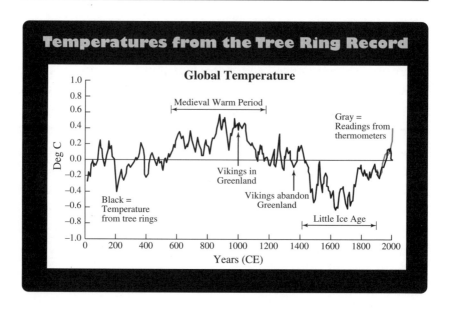

FIGURE 10.10 TEMPERATURE RECONSTRUCTION FROM TREE RING DATA

A reconstruction of past temperatures from Northern Europe over the past 2,000 years using tree-ring **data**.

These straightforward measurements on living and ancient wood thus allow us to reconstruct the temperature history over thousands of years. Figure 10.10 shows this record for Northern Europe over the past 2,000 years. The **uncertainties** in any one year's measurement (derived from dozens to hundreds of samples) can be judged from the overlapping thermometric and tree ring measurements between 1870 and 1950—it is about ±0.1°C.

As with all **proxies**, the ring width/wood density measurements carry with them certain **assumptions** about growing conditions—most simply, that nothing fundamental has changed in the way trees grow over the past 10,000 years or so. This sounds perfectly plausible, but **skepticism** is always in order. Recently, in fact, a divergence between the **proxy** temperatures from tree rings and the actual measured temperatures has suggested that something about this **assumption** is flawed. Since the early 1960s, tree ring values have been yielding flat or slightly declining temperatures while thermometers have been recording significant increases.

Further investigation of this issue, known as the "divergence problem," has shown that the effect exists only for trees from high northern latitudes; records for trees from closer to the equator continue to track the measured temperature rise. Speculation as to the cause of this discrepancy has included the effect of heat stress and/or excessive dryness in reducing tree growth, an increase in ultraviolet radiation that reaches the ground because of the thinning of the ozone layer, and the reduction in sunlight as a consequence of increased atmospheric aerosols. Note that all of these effects are, or could be, human-induced, and the fact that the effect is limited to the period after 1960 when human impacts are greatest is consistent with these hypotheses. Indeed, for the previous century, the tree ring and thermometric measurements agree very well, and at early times the various **proxies** used (from historical records of crop production to ocean temperature measurements), **data** from lake sediments, and ice core records (see figure 10.11) agree with one another, boosting confidence in the tree ring record prior to the current era.

Thus, although the individual small wiggles in figure 10.10 should not be taken too seriously, the long-term trends are worthy of attention. The period from 900 to 1100 C.E. was the warmest in the past 2,000 years with the exception of the last few decades; indeed, the period from 600 to 1200 C.E., known as the Medieval Warm Period (MWP), was all above the long-term **average**. It is not a coincidence that the peak of the MWP corresponds to the onset of massive cathedral building in Europe; a warmer climate means longer and more successful growing seasons, implying population expansion and a labor force in excess of that needed for subsistence farming. As noted in figure 10.10, the Vikings established extensive colonies in Greenland (so named because of its green pasturelands in the southern coastal regions) in the middle of this warm period. By 1350, the **mean** temperature had fallen by 0.5°C, and the Viking colonies collapsed; there was no longer enough grass harvested in the summer to keep the cattle and sheep alive in the (longer) winters, and the livestock and humans all perished.

In the ensuing century, the temperature plunged to −0.6°C below the long-term **average**, ringing in the period known as the Little Ice Age; it wasn't until 1950 that the temperature returned to its long-term **average** value (and, in the last half century, has significantly exceeded it). The seventeenth-century Dutch paintings showing snowy fields and skaters on ponds and canals record this colder climate. Today, the **mean** winter temperature in Amsterdam is a daily low of 35°F and a high of 44°F; only 10 percent of the years see the temperature fall below freezing during the period from December to February, and the **probability** of snow throughout the winter **averages** less than 15 percent.[7]

Although the tree ring record allows temperature reconstruction over the past 10,000 years, this timescale is much shorter than the periods of change in solar insolation calculated from the orbital changes noted previously. Here the permanent glaciers that cover Greenland and Antarctica provide another **proxy** with a record nearly 100 times longer and with even more detail relevant to the Earth's climate. The oxygen-isotope ratios provide a good **proxy** for temperature as described earlier. In addition, small air bubbles trapped in the ice provide direct samples of the atmospheric composition at the time the bubbles are sealed off—just as if a chemist had captured a little air and sealed it up for later analysis. Finally, records of sea salt, volcanic ash, and dust from distant deserts (Greenland ice cores show dust from the Gobi Desert in Mongolia) provide **proxies** for storminess and volcanic activity.

The glaciers are several kilometers thick, providing a record that extends back at least 800,000 years from the core extracted from Dome C in Antarctica and possibly, with future extractions, for as long as 1.5 million years. Using standard drilling technology, scientists bore into the ice and extract cores roughly five inches in diameter. Annual layers of snow accumulation are apparent even at depths of two kilometers, allowing accurate dating of the layers over hundreds of thousands of years.

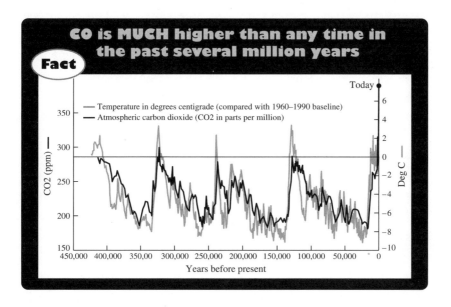

FIGURE 10.11 TEMPERATURE AND ATMOSPHERIC CARBON DIOXIDE CONCENTRATION FROM AN ICE CORE

A **time series plot** showing both the temperature as reconstructed from the ice core record and the carbon dioxide concentration from air bubbles embedded in the ice over the past 450,000 years.

The most striking **data** to emerge from the extensive work on the ice cores are summarized in figure 10.11. This figure is a graph, which, as discussed in chapter 5, includes two quantities in the same **plot** with the left and right **axes** providing the references for the two quantities—in this case temperature and atmospheric CO_2 concentration. The striking saw-tooth pattern shows five roughly equally spaced peaks in temperature over the past 425,000 years, with maxima reaching two to three degrees above today's **average** temperature, followed by steep declines to **mean** global temperatures of ~8°C (14°F) below the current value. The CO_2

concentration, ranging from 180 to 280 ppm, closely tracks the temperature fluctuations.

Figure 10.12 shows a **correlation plot** for these two variables with both linear and quadratic equations fit to the **data**. The *r value* is a remarkable 0.75, leading to a **probability** of far less than 0.1 percent that the relationship occurs by chance. However, it is always important to remember, as chapter 8 emphasizes, that **correlation is not causation**. In looking at figure 10.11, it is not at all clear whether CO_2 leads, and therefore might cause, the rise in temperature or whether the rise in temperature leads, and therefore might cause, the rise in CO_2—or, of course, if a third variable causes both to change in unison. From our discussion of how greenhouse gases work, we have a physical mechanism for how an increase in CO_2

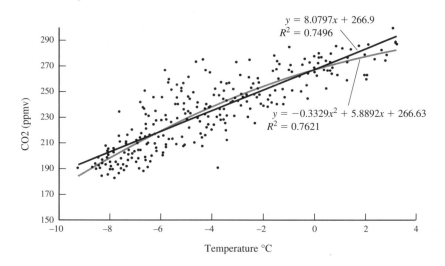

FIGURE 10.12 TEMPERATURE VS. ATMOSPHERIC CARBON DIOXIDE CONCENTRATION

A correlation plot of the carbon dioxide concentration (measured in parts per million by volume) versus temperature (in °C) showing a high-confidence correlation coefficient (R^2) from the linear fit (a quadratic fit line is also shown).

could cause an increase in temperature. However, we also know that there is a large store of CO_2 and methane (an even more potent greenhouse gas) locked up in the swampy boreal forests on the edge of the tundra across the Northern Hemisphere. If rising temperature releases even part of this reservoir, we see it could cause the rise in CO_2.

Another key feature of this **plot** is the clear periodicity of the glacial and interglacial periods; it is almost precisely that predicted by the changes in the ellipticity of the Earth's orbit discussed earlier (slightly over 100,000 years). In some cycles, it is even possible to see signatures of the 41,000- and 23,000-year cycles of the orbital parameters (see peaks separated by ~40,000 years in the most recent cycle and by ~25,000 years in the previous one). This is the justification for my earlier statement that these orbital effects have a major influence on the Earth's climate.

The final **datum** worthy of note in figure 10.11 is the point that represents the CO_2 concentration today. It is clearly much higher (>40 percent) than the level seen at any time in the last 425,000 years. The deepest ice core we have extends this statement by another factor of two to 800,000 years. Various geochemical techniques have allowed a reconstruction of the CO_2 concentration over hundreds of millions of years and imply the current level of 400 ppm has not been exceeded for twenty-two million years. In the remote past, however, levels were much higher, reaching 6,000 to 7,000 ppm between 200 million and 500 million years ago and remaining above 1,000 ppm until shortly after the demise of the dinosaurs sixty-five million years ago.

SCENE 5

Given a healthy dose of physics and a huge collection of **data**, it is possible to construct computer-based, **numerical models** of the Earth's climate. These are called general circulation models (GCMs) or, sometimes, global climate models, and they seek to capture the interactions of the atmosphere, oceans, ice, and land while including external factors such as solar output and orbital changes in the calculations. They can be tested against paleoclimate **data** to see whether they give consistent results and used to predict the future given various assumed changes applied to the energy-in = energy-out equation.

An example of **postdiction** is shown in figure 10.13. The **data**, represented by the black lines (which are identical in the two plots), are global temperature measurements since 1900. The gray lines represent the **mean** of about a dozen different GCM predictions. The bottom panel shows the predictions of the **models** when only natural phenomena are

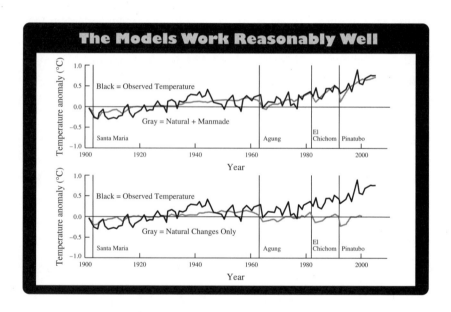

FIGURE 10.13 MEASURED GLOBAL TEMPERATURES VS. MODEL PREDICTIONS

Global temperature records over the last century (plotted in back) compared with **models** (plotted in gray) that both exclude (lower graph) and include (upper graph) human contributions to climate change. The dates of major volcanic eruptions are marked.

included (e.g., note the dips for every major volcanic eruption). They parallel the actual **data** well up until about 1960, at which point they begin to diverge, ending up short of the observed temperature today by roughly 0.8°C. The top panel shows the same **data** overlaid with the **mean** GCM predictions in gray when both natural and manmade effects (predominantly the rise in greenhouse gases) are included. The agreement is impressive.

Although the discrepancy between the curves that include and exclude **anthropogenic** effects is striking, the agreement in the top panel should be less so to an appropriately **skeptical** scientific mind. The GCMs typically have millions of lines of computer code running on the fastest supercomputers in the world. They have dozens, if not hundreds, of input parameters. It is unsurprising then that, after tuning all the parameters available, the **models** can match existing **data** and that they largely agree with one another when predicting the future. If your **model** gave a **postdiction** widely discrepant from the **data,** you'd adjust some parameters to make it work. If it gave a prediction wildly different from that of the consensus of other GCMs, would you bravely publish (or try to publish) it, or would you look to tweak some more parameters? **Publication bias** must be considered when we find that all of the **models** agree on the future. Indeed, according to the latest report from the Intergovernmental Panel on Climate Change, the rate of global warming between 1998 and 2012 was overpredicted by 111 of the 114 **models** tested.[8] If the **models** were all completely independent of one another with only **random errors** leading to discrepancies from reality, we'd expect roughly fifty-seven to be high and fifty-seven to be low. The **probability** of getting 111 to all be high is less than 5×10^{-7}. This suggests instead a **systematic** bias that is shared by the vast majority of the **models**—or effects in the real climate system that the models all fail to take into account.

With that important caveat in mind, let's see what the **models** predict for the remainder of this century. Figure 10.14 shows the period

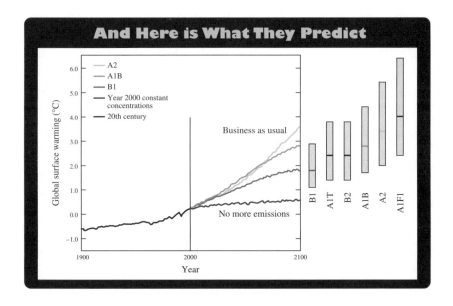

FIGURE 10.14 GLOBAL TEMPERATURE MODEL PREDICTIONS

Predictions for global temperature rise over the next century as a function of various assumptions about greenhouse gas emissions. The different scenarios (labeled A2, A1B, etc.) are those included in the report of the Intergovernmental Panel on Climate Change (www.ipcc.ch).

from 1900 to 2100 with a smoothed version of the temperature **data** from figure 10.13 up to the year 2000, followed by the **model** predictions for this century. The principal variable is the different scenarios assumed for the input of greenhouse gases to the atmosphere: from zero emission starting in 2000 (already exceeded), which predicts an additional temperature rise of 0.6°C by 2100, to the "business-as-usual" (meaning emissions continuing at their current level), which leads to a rise of nearly 4°C (7°F).

SCENE 6

In addition to simply predicting **average** global temperature, GCMs can be harnessed to explore other aspects of the climate system. For example, the eight million square kilometers of sea ice that blanketed the Arctic ocean at the end of the Arctic summer in 1980 is a crucial component of the global climate system. As noted earlier, declining ice coverage can trigger an important **feedback loop** in that the less reflective ice there is, the more energy is absorbed, and the warmer it gets, the less reflective ice there is. Figure 10.15 shows a range of **models** (light gray lines), all of

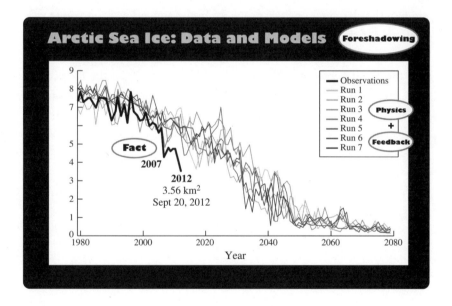

FIGURE 10.15 MEASURED ARCTIC SEA-ICE COVER VS. MODEL PREDICTIONS

A comparison of **models** (various light gray lines) predicting the shrinking of Arctic sea ice with the **data** (thick black line) over the past thirty-five years.

which predict rapidly declining ice coverage such that the Arctic could be ice-free at the end of summer in 2080. The **data** are shown in black. The measurements followed the predictions pretty well until 2005. By the summer of 2007, however, September ice cover had dropped close to four million square kilometers (half that seen in 1980) and 30 percent below the **model** predictions. In 2012, another plunge brought the coverage on September 20 to 3.56 million square kilometers, almost a factor of two below the predictions. **Feedback** is, of course, included in the **models**, but it is clear they do not predict the rapid observed decline that could lead to an ice-free North by 2050 or before.

This is one example of the foreshadowing I referred to at the beginning of this chapter. Another example is the state of the Greenland ice cap. Problem 6 of the chapter 4 practice problems in the appendix allows you to calculate that if the entire Greenland glacier melted, global sea level will rise by about twenty-three feet. That's not going to happen anytime soon. However, the air temperature over Greenland has risen 4°F in the last twenty years, greater than the amount predicted from the **models**. In addition, the glacial flow rate in the regions where the ice meets the sea and where iceberg calving occurs has doubled in five years—that was also not predicted. At the other end of the world, however, the Antarctic sea ice is thickening somewhat; again, this was not predicted by the **models**. The recent news about the now irreversible collapse of the West Antarctic ice sheet is yet another example of foreshadowing events that demand our attention.

SCENE 7

Before ending my lecture, I like to point out a few fictions that are repeated so often they become accepted as fact when, in reality, they run counter to the best evidence available and serve only to confuse the discussion or, worse yet, provide an easy target for ideologically driven critics of climate science. One example is the cause of sea-level rise, observed to be a little over three millimeters per year. In addition to the physically impossible notion that melting Arctic sea ice is responsible (floating ice does *not* increase the volume of water when it melts), the oft-repeated explanation is that melting glaciers, such as the big ones in Greenland, are adding water to the oceans. Although this does occur, Greenland's contribution is estimated to be 0.4 ± 0.1 mm per year or about 10 percent of the total. In fact, the real explanation for most of the sea-level rise is that the oceans are expanding as the temperature rises (see box 10.6). North Carolina law notwithstanding, a continuing rise in temperature will inevitably

BOX 10.6 ESTIMATING SEA-LEVEL RISE THAT RESULTS FROM WATER'S THERMAL EXPANSION

The amount by which a substance expands in response to a change in temperature is called its coefficient of thermal expansion. For water, this quantity is 0.000214 per degree Celsius. Over the past fifty years (see figure 10.13), the **mean** global temperature has risen about 0.7°C. The **mean** depth of the ocean is 3.8 km. If one pictures the ocean as existing in a (funny-shaped) beaker with straight sides, heating the water simply means the level of water in the beaker will rise. The real ocean, of course, is not so rigidly contained; if sea levels rise enough, it will flood land, thus expanding its surface area and leading to a smaller overall rise. However, the beaker approximation is good enough for a **back-of-the-envelope** calculation in which we are discussing raising levels by millimeters per year.

If the entire ocean immediately came to equilibrium with the temperature of the air, we can calculate the expected level of rise quite simply as $0.000124 \times 3.8 \text{ km} \times 10^6 \text{ mm/km} \times 0.7°C/50 \text{ years} = 6.6 \text{ mm/year}$.

This is slightly more than twice the total observed rise and at least three times the amount normally attributed to thermal expansion. The principal reason is that the ocean does not come to equilibrium with the air instantaneously. In fact, the deep ocean waters take centuries to

---

(continued)

respond to changing air temperature, with only the surface layers heating quickly through their contact with the atmosphere. Nonetheless, this estimate shows that thermal expansion is an essential component to consider when predicting future sea levels. Today, roughly two-thirds of the rise we observe is likely a consequence of this inevitable physical process; over the next century, with a four-degree Celsius increase in global temperature, the total effect could be measured in meters, not millimeters.

---

lead to a rise in sea level independent of the somewhat poorly understood details of how Greenland and Antarctic ice are melting.

Another one of my favorite fictions to skewer is the claim that rising temperatures lead to stronger hurricanes and other storms. *An Inconvenient Truth* draws a direct connection between global warming and the destruction that Hurricane Katrina wrought on New Orleans. Attributing any particular storm to climate change is, first of all, ludicrous. It completely misses the important distinction between climate and weather. Furthermore, the **data** from the National Hurricane Center (whose breathless interviewees cultivate this fiction on cable news networks throughout each hurricane season) are unambiguous in revealing the falsity of this claim.

The period from 1930 to 1970 was one of relative stability in the global temperature, with a **mean** value close to the long-term **average** (see figure 10.13). The period from 1970 to 2012 has seen the **mean** global temperature rise by about 0.7°C. What are the hurricane **statistics** over these two intervals? From 1930 through 1969, there were 235 Atlantic hurricanes; if we assume the number of hurricanes each year, or decade, or forty-year period is **randomly distributed**, the **uncertainty** in this number from a **statistical** perspective is $\sqrt{235}$ = 15, so 235 ± 15. In the rapidly warming forty-year interval from 1970 to 2009, there were 239 ± 15 hurricanes for a difference of 4 ± 22, not even one-fifth of a **standard deviation** different from zero. Furthermore, if we take only "major hurricanes" (those defined on the Saffir–Simpson scale as Category 3 or higher) during the interval from 1930 to 1969, there were 107 major storms compared with 93 from 1970 to 2009 (a difference of –14 ± 14—one **standard deviation**, implying the **odds** are only 16 percent that we would get these results by chance if *colder* periods were not more prone to storms).

Finally, there are the long-term **data** from the ice cores. As figure 10.16 shows, the periods in which storms carry dust from distant lands onto the Greenland ice cap reach a sharp maximum when the global temperature is *lowest* and fall to zero (down by at least a factor of twenty-five) during the warm interglacial periods. In summary, there is no evidence that tropical cyclones have increased in frequency or intensity in step with the rapid global warming of the last fifty years. This mantra is fictional; some of the illogical arguments and claims that support it can be found in the next chapter.

SCENE 8

Consistent with my **skeptical**, scientific, detached perspective on the issue of climate change, I do not draw a conclusion for my audience. The goal of my presentation—and the goal of this book—is to provide you with the tools for coming to informed and independent judgments about important issues such as this one. I do share, however, a comparison of the list of things my audiences are conditioned by the media to worry about and the list of things I actually worry about. In the former category I include stronger and more frequent hurricanes, rising sea levels, hotter summers and more deaths from heat stroke, and the demise of polar bears.

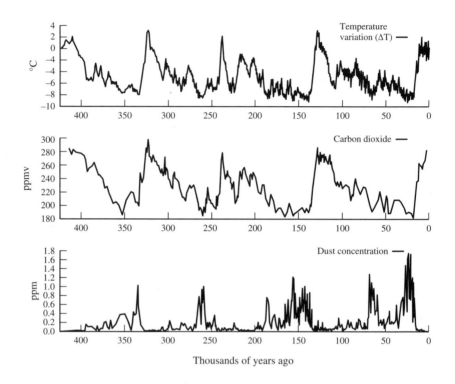

FIGURE 10.16 TEMPERATURE, CARBON DIOXIDE CONCENTRATION, AND WIND-BLOWN DUST FROM THE ICE CORE RECORD

The ice core record over 400,000 years showing the temperature departures from the twentieth century average (in °C), the carbon dioxide concentration (in parts per million by volume), and the amount of wind-blown dust (in parts per million) as a **proxy** for storminess.

The first issue I have dealt with here and will revisit in chapter 11. The second issue is a concern if you live in Bangladesh or in some coastal areas of Florida, but at three millimeters per year, it is not going to transform the world economy or shake up the geopolitical order in the next few decades. From 1999 through 2010, the number of heat-related deaths in the United States **averaged** 618 per year, with a variation of less than a factor of two in any one year and no **secular** trend. For comparison, the number of traffic-related deaths in 2010 was 32,788, and the number of deaths involving firearms was 31,672. And although I like polar bears as much as anyone, the things I worry about are the spread of tropical disease vectors and emerging zoonotic diseases, the collapse of biodiversity, CO_2 from boreal bogs, and freshwater exhaustion.

In 2007, the mosquito-borne dengue fever-like virus chikungunya infected more than 100 people in Northern Italy; this will not be the last time tropical diseases reach temperate climates. The current extinction rate, although measured over only a very small slice of geologic time, is estimated to be between ten and 1,000 times the average rate experienced over the past half-billion years. As noted earlier, there is an enormous storehouse of greenhouse gases in the boreal bogs of the Northern Hemisphere (greater than the total greenhouse gas contribution from 200 years of burning fossil fuels), and its release could drastically affect atmospheric composition. Finally, the reckless drawdown of water from deep aquifers and the exhaustion of ground and surface water in many major agricultural areas of the world is, in my view, likely to lead to a crisis in freshwater availability within twenty years.

Neither set of worries is easily dismissed, but the latter set is here and now, whereas the former includes myths and concerns that involve much longer timescales.

Before ending my lecture with three aphorisms, I ask the audience to consider the key evidence presented. Very briefly, I regard this evidence as follows:

- ▶ Many factors affect the Earth's climate, and all of them change with time.
- ▶ Of these many factors, some lie permanently beyond our control.
- ▶ We are altering in a significant way one of these factors—the chemical composition of the atmosphere—as a direct consequence of human activity.

- Atmospheric CO_2 concentration is now higher than at any time in the last 450,000 years and, most probably, in the last twenty million years.
- Well-established physical mechanisms exist that suggest this increase in greenhouse gases will lead to higher global temperatures.
- The **mean** global temperature has risen ~$0.7°C$ in the last fifty years and is now higher than at any time in the last several thousand years at least.
- We have **models** that predict a much greater rise in temperature (and other ancillary impacts) if we continue operating as we are now.
- These models are bedeviled by many **feedback** loops and other uncertainties but are good at reproducing past changes of climate.

And now for my three aphorisms:

1. This is not the first time living creatures have fundamentally altered the Earth (cyanobacteria completely changed the atmospheric composition billions of years ago)—it is just the first time such a creature could decide whether or not to do so.
2. This is not the first time the Earth's climate has changed (40 million years ago, there were palm trees in Greenland)—it is just the first time the change may be within a species' control.
3. This is not the first time the Earth's future is uncertain (the dinosaurs didn't know about the incoming asteroid)—it is just the first time that a species finds "the future" a concept worth contemplating.

My final slide is a cartoon from the *Christian Science Monitor*. It shows two theater marquees. One theatre is showing "*An Inconvenient Truth*" and the ticket taker sits alone with no customers. The adjacent theater is showing "*A Reassuring Lie*"; the line to get in snakes around the block. As you might infer from the foregoing, I don't want to stand in either line. I want a third option: a rational, dispassionate analysis of the situation that sweeps away the mountains of misinformation born of ignorance, ideology, and self-interest and builds toward understanding using scientific habits of mind.

11
WHAT ISN'T SCIENCE

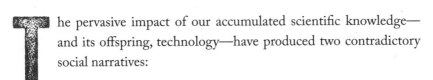he pervasive impact of our accumulated scientific knowledge—
and its offspring, technology—have produced two contradictory
social narratives:

     1. Science is enormously powerful, and to have similar stature in
both academic and real-world settings, a discipline, procedure, or activity must
acquire the trappings of science.

     2. Science has produced enormous negative consequences in many spheres of
life, and only the rejection of science and a return to aboriginal and/or personal
ways of knowing can save us.

The only feature these two narratives share is the magnitude of the impact
of science on individuals, society, and the natural world. Of this there can
be little doubt. In virtually every sphere of human activity and in every
corner of the planet, the footprint of the scientific knowledge accumulated over the last four centuries is evident:

> Communication: From quill pens to e-mails, texts, and cell phones . . . to
>   the National Security Agency (NSA)
> Transportation: From walking to cars, planes, jets, and . . . air pollution

Medicine: From leeches to biotics, anesthesia, microsurgery, and . . . antibiotic resistance

Anatomy: From dissecting cadavers to x-rays, computerized axial tomography scans, functional magnetic resonance imaging, and . . . a black market in organ donors

Materials: From blacksmiths to composites, superconductors, carbon nanotubes, and . . . chemical waste

Computation: From abacus to i-Pads, laptops, supercomputers, and . . . cybercrime

Military: From sabers to nuclear weapons, spy satellites, drones, and, well . . . forget it—they're already all bad

Energy: From wood to fossil fuel, hydroelectric and nuclear power, and . . . climate change

Entertainment: From court jesters to i-Pods, computer-generated imagery movies, *SimCity*, and . . . *Grand Theft Auto V*

Education: From monastic libraries to tablets, Google, MOOCs, and . . . MOOCs

Food production: From oxen to synthetic fertilizer, miracle rice, genetically modified organisms, and . . . algal blooms produced by phosphate runoff

Human relationships: From letters to phones, Skype, Facebook, and . . . cyberstalkers

Astronomy: From astrolabes to ten-meter telescopes, the Hubble Space Telescope, intercontinental radio arrays, and, well . . . there's nothing bad about astronomy!

Many people would be happy to live without the NSA, climate change, chemical waste, and even *Grand Theft Auto V*. Many fewer would be willing to see 50 percent of their children die before the age of 15 (UK data from the early 1600s[1]), have a life expectancy of twenty-five (as in the colony of Virginia during the same time period[2]), never hear music, and walk everywhere they needed to go.

The goal here, however, is not to debate the moral standing of scientific knowledge or its acquisition. Rather, in this chapter I will enumerate some of the activities that, adopting social narrative 1 as outlined previously, claim to be, or are confused with, science and that, in my view, are not science. In chapter 12 I will discuss examples of social narrative 2 and the perils I see in adopting such an approach given our current global circumstances.

EXAMPLES OF PSEUDOSCIENCE

Philosophers and historians of science do not like the term pseudoscience because it harkens back to the "demarcation problem" discussed by Karl Popper (see chapter 2). Various alternatives have been suggested: voodoo science, pathological science, nonscience, etc., but just as I am happy with Popper's definition that science is centered on falsifiability, I am comfortable with the word pseudoscience and will offer examples here.

Good science is done by those who are skeptical, empirical, disinterested, scrupulous in their intellectual honesty, and open to whatever outcomes Nature produces. Pseudoscientists, by contrast, are believers, free of skepticism, interested (often financially), intellectually dishonest, and dogmatic. The demarcation is pretty clear to me. That is not to say there are no fuzzy areas (I'll briefly mention a couple) or that there is no bad science (there's lots of bad science unfortunately, and I will provide examples). But I see no reason not to label pseudoscience for precisely what it is—false science.

ASTROLOGY

Since I am an astrophysicist, you will be unsurprised to find that my first example of pseudoscience is astrology. If you perform a Google search for astrology, the first six sites that come up are, unsurprisingly, places to get your horoscope—or the latest breaking news in astrology.

I am writing this chapter two days after the actor Robin Williams committed suicide. One of the news flashes for astrologers and their believers, found on the appropriately named website Beliefnet.com, reads exactly as follows (bold in original):

> **Robin Williams (born July 21, 1951, 1:34 PM Chicago Illinois) was arguably the funniest human of his generation.** He's best known as a comedian, and launched to fame in the 1980s in the sitcom "Mork and Mindy." But his talents were too great to be contained by a sitcom, and before long he was starring in major films and was a leading force behind "Comic Relief USA" which raised millions for the homeless. **We can look at Robin Williams' birth chart and see the signs of potential depression:** Moon in sensitive Pisces opposite Venus in Virgo. A Moon in Pisces is sensitive enough to life's discomforts as it is, without any oppositions. The Moon's ruler was Neptune, and Robin's Neptune was in the 12th house. This is always the mark of someone who could stand to have some form of spiritual retreat in his life. It seems unlikely that the nature of his busy Hollywood lifestyle would allow for that sort of thing a lot, although he did end up in rehab a few times. The potential for something alone does not guarantee it will definitely happen. **There was more to this suicide then [*sic*] simply indicators of depression in the birth chart.** There were difficult transits happening when he died . . . that could trigger a suicide: transiting Pluto was closely opposing Robin's Mars-Uranus conjunction at the time of his death. A Pluto transit opposite either planet is going to be a challenge, and it can be argued that suicide is aggression (Mars) turned inward. Saturn's recent passage over Robin's Ascendant and close square to his natal Pluto undoubtedly weren't helping either. Under the best circumstances, any one of these transits could be the sort of thing that would indicate a significant emotional crisis. And perhaps all of them

put together were the final straw for Robin Williams, who had always struggled with depression. **Anyone getting tough transits from both Saturn and Pluto is going to have a hard time.**[3]

I suspect many of my readers will find this as ludicrous as I do. It is a mistake, however, to dismiss it as having no impact on societal attitudes and public policy. Consider the following story from the *Telegraph*:

A Conservative member of the Commons health select committee has said he is "absolutely convinced" of the benefits of astrology and called for it to be incorporated into medicine.

David Tredinnick said he had spent 20 years studying astrology and healthcare and said it had a "proven track record" guiding people through their lives. The MP, a member of both the Commons health select committee and the science and technology committee, is a keen advocate of complementary therapies, and chairman of a Government working group on herbal medicine. On Friday he said more should be done to raise awareness among patients and healthcare professionals of the benefits of astrology.[4]

The article goes on to report that Tredinnick has labeled those who have criticized his position as "ignorant bullies" and that he has cast horoscopes for a number of members of parliament (MPs) who had come to him for that service. The article also noted, however, that he was forced to pay back 755 pounds in the MPs' expense scandal a few years earlier because the investigators felt that astrological software was an inadmissible business expense for an MP. Apparently they are more concerned in the United Kingdom about how government expense accounts are used than how healthcare is managed.

This is not to mention Nancy Reagan's personal astrologer who is reported to have strongly influenced President Reagan's schedule and decision-making.

There have been a number of scientific studies of astrology,[5] and all of them have found no evidence whatsoever for the predictive power of any astrological system (indeed, an important fact to note is that there are a number of independent and radically different astrological systems, all of which claim predictive power). No plausible physical mechanism exists to explain how celestial bodies influence events on Earth or how targeting of individuals is possible. One common pseudoscience claim is that the Moon and the Sun raise tides on Earth, and since the human body is 60 percent water, that's an obvious physical mechanism. The first problem with this pseudo-explanation is that it betrays fundamental scientific ignorance: tidal forces do not just act on water but on everything with mass. The second issue is that one can easily calculate the gravitational force between two bodies, and the force of the obstetrician on a newborn at the time of delivery is greater than that of Mars—yet somehow we have escaped a "science" of obstetrology.

HOMEOPATHY

I can't leave British eccentricity on the matter of healthcare without addressing homeopathy, a "therapeutic" treatment touted by Prince Charles, heir to the British throne, and fully covered by the UK's National Health Service. This technique, invented by the German physician Samuel Hahnemann in 1796, consists of providing the patient with dilutions of putatively active ingredients that supposedly would trigger the patient's symptoms in a healthy human and can thus cure disease on a theory called the "law of similars." The dilutions are carried out such that often not a single molecule of the putatively active ingredient exists in the dose given to the patient.

The fact that homeopathy is completely inconsistent with all current biological knowledge and is physically and chemically impossible leaves true believers undeterred. According to the National Center for Complementary and Alternative Medicine (NCCAM)—a multibillion dollar boondoggle funded as part of the National Institutes of Health by representatives and senators of Tredinnick's ilk—U.S. consumers spent over $3 billion on homeopathic medicine in 2007,[6] obtaining distilled water from which the last trace of such invaluable ingredients as crushed whole bees, red onions, and white arsenic were originally dissolved (a terrible waste of bees in my view).

One might argue that the 3.9 million U.S. adults who opted for homeopathic cures in 2007 were entitled to make their own healthcare decisions, and if they chose to opt for fantasy treatments instead of science-based medicine, that's acceptable. But the 910,000 children who received homeopathic treatments in that same year are, to me, more disturbing. The fact that (1) the Food and Drug Administration allows homeopathic remedies "to be marketed without Agency preapproval" and "does not evaluate the remedies for safety or effectiveness,"[7] (2) many states license homeopathic practitioners, lending them the imprimatur of government, and (3) some health insurance plans cover the cost of these treatments are all evidence of the hold that arrant nonsense, masquerading as science, has on our social institutions.

ACUPUNCTURE

A major component of traditional Chinese medicine, acupuncture has achieved acceptance in Western societies beyond that of many other alternative medicine treatments. It is also pure pseudoscience. It relies on using needles inserted into the body to redirect the flow of qi—the life force energy—along the meridians of the body. The fact that there is absolutely no evidence for the existence of qi, or of the meridians along which it supposedly flows, is of no consequence to believers, of course, since once one has belief, evidence is superfluous.

Over 3,000 (yes, three thousand!) studies of acupuncture have been published, and the conclusion is clear—there is no measurable therapeutic effect of acupuncture for any condition that cannot be attributed to the placebo effect. Many of these studies have shown that sham acupuncture (when patients are told they are undergoing treatment but in fact nothing is being done) works just as well as "real" acupuncture (leaving aside the question of what "real" means in this context). Unsurprisingly, the Mayo Clinic cites "evidence that acupuncture works best in people who expect it to work."[8]

Again, one might argue that participation in a medical procedure that has no benefit beyond the placebo effect is a harmless adult activity. But acupuncture isn't risk-free. In the largest systematic study to date, 229,230 patients undergoing acupuncture therapy were followed. A total of 19,726 of the patients (8.6 percent) experienced an adverse effect, of which 4,963 (2.2 percent) required (real) medical treatment.[9] Between 2000 and 2009, five deaths directly attributable to acupuncture sessions were reported in the United States.[10]

Acupuncture, homeopathy, and a host of other popular pseudoscientific medical treatments are neither risk- nor cost-free. A 2007 NCCAM survey found that Americans spent $33.9 billion on these sham treatments, much of it out of pocket because many health insurance plans do not cover such treatments (although many do). This injustice outraged U.S. Representative Maurice Hinchey, a New York Democrat, and prompted him to introduce legislation in 2011 to add acupuncture coverage to Medicare and to federal employees' health insurance plans. Hinchey's spokesperson introduced this initiative with a flat falsehood:

> Even though the National Institutes of Health has found acupuncture to be an effective treatment for a range of serious diseases and conditions, 52 million Medicare beneficiaries and federal employees have no guaranteed access to the treatment.[11]

The spokesperson went on to describe that the purpose of the Federal Acupuncture Coverage Act is to "ensure that those in need of care aren't forced to choose more expensive, invasive and risky treatments with a long list of side effects simply because acupuncture isn't covered."

The media duly reported this important government initiative and solicited expert commentary on the wisdom of this proposed new waste of taxpayer money by asking acupuncture practitioners for comment. Despite annual reintroductions by various members of congress from both sides of the aisle, as of 2015 (you may be relieved to know), the act has not become law.

PARAPSYCHOLOGY

Astrology can claim a heritage of several thousand years; studies of the paranormal have only failed to produce a single scientifically validated result for about 150 years. The host of pseudoscientific activities usually grouped under this general category include reincarnation, telekinesis, precognition, telepathy, out-of-body experiences, and various other super-natural phenomena, sometimes called psi or extrasensory perception. As with the other pseudosciences discussed earlier, every careful study of these phenomena has shown no reason whatsoever to take any of them seriously.

Nonetheless, every so often, an article will appear that keeps the psi flame burning. In 2011, retired Cornell psychology professor Daryl Bem published a study on precognition[12] that was the latest result to achieve wide media attention—after all, the author was an Ivy League professor (that fatal appeal to authority), and the article was published in the peer-reviewed *Journal of Personality and Social Psychology* (that fatal appeal to infallibility in the system of science that no skeptical scientist has). Since it was a carefully written and peer-reviewed article, however, it did con-form to many standard scientific practices. This allowed experts to address the statistical treatment of the data (which was flawed) and allowed other experimenters to attempt to replicate the results—which they tried to do and, using thousands of subjects, found no results whatsoever to confirm the original claim.

In the course of the media storm over Bem's original article, I was asked to write a column for the *New York Times* on the issue of peer review fallibility.[13] Noting that peer review was a human activity and that all human endeavors entail fallibility, I nonetheless strongly defended the peer-review process as part of the powerful system of self-correction embedded in the practice of science. For this, I was labeled a "science jihadist" by one of the many commentators on my column. It is a mantle I wear proudly.

Although many people dismiss scientific studies as ill-equipped to explore psi since, by definition, psi phenomena lie outside the sphere of normal science, few people dismiss money as a motivator. Fifty years ago, the magician James Randi offered a prize of $1,000 to anyone who could demonstrate any paranormal phenomenon under mutually agreeable conditions. With support from skeptics such as Johnny Carson, the prize was raised to $1 million. Over 1,000 people have applied to be tested in an attempt to win the prize. Zero entrants have succeeded, and the prize remains unclaimed.

EXAMPLES OF SCIENCE "CONTROVERSIES"

CLIMATE CHANGE

There are many people for whom the facts of climate change are inconvenient—these facts threaten their livelihood, their profits, or, apparently in Senator's James Inhofe's case (see chapter 10), their religious beliefs. (I say "apparently" because it is worth noting that Inhofe represents the state with the fourth largest oil production per capita in the country, and one cannot completely dismiss this as at least an ancillary motivation for his position on the subject.) Jobs, stock prices, and religious beliefs are all matters worthy of consideration in forming public policy. They do not, however, either replace or trump scientific facts.

In debates on topics such as climate change, when the data fail completely to support an interest or a prejudice, all that's required is to come up with alternative data or plausible-sounding theories (recall from chapter 2 Popper's condition of falsifiability of any theory worthy of the name). This applies both to climate change promoters and deniers. On the topic of hurricanes, for instance, we find an example of the "new theories" category in the mantra that storms draw their energy from warm water, and the warmer the water, the stronger the storms will be.

While it is true that warm water contains more energy than cold water, if this simplistic statement were true, there would have been a significant increase in powerful hurricanes over the past thirty years, during which the mean global sea surface temperature has risen more than one degree Fahrenheit. Instead, as demonstrated in chapter 10, the number of strong storms has decreased during this period.

Or we can change the data we look at completely and, for example, quote property damage estimates from storms. Unsurprisingly, these have risen dramatically since the 1930s. Is that evidence of stronger storms? Of course not. It is evidence of rampant coastal construction underwritten by an economically insane federal flood insurance program.

The total losses estimated for each decade from the 1950s through the first decade of this century are, respectively (in billions of dollars), $2.3, $5.8, $9.3, $26, $62, and $237. Does this imply that the destructiveness of hurricanes has increased more than 100-fold in the last fifty years? No. But the enlightened policy of North Carolina described early on in chapter 10 that outlaws using science to make policy assures that property damage from storms will continue its rapid rise. This outcome, however, has absolutely nothing to do with the increasing strength of storms—or with science.

CREATION SCIENCE

In 1654, James Ussher, Archbishop of Armagh, published a biblical chronology purporting to show that the creation of the Earth occurred in the early evening of October 23, 4004 B.C. From a scientific perspective, this moment of creation and that postulated by the five-second-agoers described in chapter 9 are equally plausible. There is nothing fundamentally wrong with believing Ussher's date (except that one would have to ignore 400 years of human intellectual history), but this belief is not remotely scientific. It does not, then, belong in science textbooks or science classrooms. That is the essence of the court case with which I opened chapter 9 and the many cases that have occurred since. Adopting the trappings of science does not make something science.

In the cases both of climate change and of creationism, the media, our politicians, and the public at large follow an instinct that sounds plausible but is logically flawed. Using the analogy of political debates, the argument goes that if there are two sides to an issue, they should get equal time to present their views. I would agree with this concept if the issue to be debated were whether a scientific view or an anti-scientific view were superior—those are two separate views, and one could argue they deserve the opportunity to state their cases. But there *aren't* two sides to the *scientific* question of whether humans are changing the composition of Earth's atmosphere, and there *aren't* two sides to a scientific discussion of how life evolved.

The standard rhetoric (adopted interestingly by climate change deniers and creationists alike) is to "teach the controversy." But there is no controversy. There is much vigorous and healthy debate within the scientific community about the details of climate models and their predictions, the calibration of proxies, the best tools to use to further the science, etc., but there is no controversy about whether human activity is changing the composition of the Earth's atmosphere and oceans and, as a consequence, altering the climate. Likewise, there are serious disagreements among evolutionary biologists about how much weight to put on genetic rather than morphological changes as markers of evolution and about the detailed timing of speciation, etc., but there is no debate anywhere in the biological sciences about the basic picture of evolutionary change. There is a scientific understanding of evolution and a religious belief about creation. These are distinct approaches to the question as to how humans came to dominate the Earth, and it is possible to have debates about the value or utility or "truthfulness" of science versus religion (I have participated in a few myself), but that is not a scientific debate, and there is no scientific controversy.

The fundamental logical flaw in the teach-the-controversy view arises from the fact that science is not democratic—only Nature gets a vote, and her results are incontestable.

EXAMPLES OF BAD SCIENCE

Vested corporate interests and the politicians who serve them, biblical literalists, and misguided consumers are not the only generators and consumers of bad science. People who wear the mantle of science but who have had little experience in scientific research—most notably medical clinicians—and full-fledged research scientists themselves also produce bad science.

SMALL CLINICAL AND PRECLINICAL STUDIES

The largest amount of bad science being produced today—and probably that with the largest impact—comes from small clinical and preclinical studies (e.g., the use of mouse models or cancer cell cultures) in hospitals, universities, and medical schools. The quality of this work is compromised by small sample sizes, poor experimental designs, and weak statistical reasoning; add a healthy dose of publication bias and the results are truly appalling.

Dr. C. Glenn Begley was head of global cancer research at the biotech company Amgen for a decade. Over that period, he selected fifty-three articles with supposedly "landmark" status (major results, trumpeted as such by the journals that published them and the media that reported on them) and assembled an Amgen team to attempt to reproduce the results. They found that forty-seven of the fifty-three studies (89 percent!) were irreproducible—the results were simply wrong.[14] More disturbingly, when the Amgen team contacted the papers' original authors in an attempt to understand the discrepancies, some would cooperate only on the condition that the Amgen scientists sign confidentiality agreements forbidding them to disclose any data that cast doubt on the original findings, thus assuring that their phony results would stand in perpetuity in the scientific literature.

A year earlier, a team at the big pharmaceutical firm Bayer AG attempted to replicate forty-seven different cancer study results and found more than three-quarters of them to be irreproducible. The article that reported this result was appropriately titled "Believe It or Not."[15]

The vast majority of these articles are not examples of deliberate fraud, however. As Begley and Lee Ellis state:

> The academic system and peer-review process tolerates and perhaps even inadvertently encourages such conduct. To obtain funding, a job, promotion or tenure, researchers need a strong publication record, often including a first-authored high-impact publication. Journal editors, reviewers and grant-review committees often look for a scientific finding that is simple, clear and complete—a "perfect" story. It is therefore tempting for investigators to submit selected data sets for publication, or even to massage data to fit the underlying hypothesis.[16]

The conclusion is inescapable: this work represents bad science. It fails the values of skepticism and disinterest, as well as the authors' responsibility to attempt to falsify their results. Indeed, I would go so far as to say it is not science, and the journal editors and funding agencies and faculty committees that allow it to persist are simply feeding the antiscientific and anti-intellectual tendencies of society to the peril of us all.

FRAUD

The systematic mess discussed above is, we are told, largely a matter of individual misunderstanding of what true scientific habits of mind entail and the social pressures of the academic system. Worse still is outright fraud. Often attributed to the same academic pressures such as the need for jobs, grants, and promotion to tenure, there is little doubt that the number of incidents of outright fraudulent publications is growing.

A study by Fang et al.[17] in the *Proceedings of the National Academy of Sciences* examined all 2,047 papers that had been retracted after being published and listed in the National Library of Medicine's MEDLINE database before May 3, 2012. They found 436 cases of errors, some caught by the original authors and some caught by others; given that the database contained roughly twenty million publications at the time, this is an error rate most occupations would envy (although there is no reason to think that every error has been caught, given the reproducibility studies cited previously). More disturbingly, they found 201 cases of plagiarism and 291 cases of duplicate publications (sometimes called self-plagiarism); both instances represent unacceptable departures from ethical norms in the academic world (scientific or otherwise). The largest fraction of papers retracted, however—888 (43.4 percent)—were because of outright fraud or suspected fraud. Based on these data, the authors claimed this percentage has risen ten-fold since 1975.

A separate study,[18] also published in 2012, went beyond the biomedical sciences to include forty-two major bibliographic databases across academic fields. The authors of this study documented 4,449 retracted articles originally published between 1928 and 2011.[19] The headline-grabbing nugget from the study's abstract was that the number of articles retracted per year had increased by a factor of 19.06 (a number clearly reported to too many significant digits) between 2001 and 2010;[20] the increase dropped to a factor of eleven when repeat offenders and the overall growth in papers published were included in the calculation.

While all incidents of scientific fraud are abhorrent, none of the media coverage noted the tiny fraction of papers involved and the even smaller fraction of scientists. Repeat offenders are responsible for a significant fraction of retractions; for example, most of the 123 papers retracted from *Acta Crytallographica Section E* were attributable to two authors. And of the 76,644 articles published in the *Journal of the American Chemical Society* between 1980 and 2011, twenty-four (0.03 percent) were retracted; the numbers for *Applied Physics Letters* were even better: 83,838 papers, of which fifteen (0.018 percent) were withdrawn (not all for fraud). If all our social enterprises had a failure rate of less than 0.1 percent and upheld ethical norms 99.982 percent of the time, it is likely the world would be a better place.

The impact of scientific fraud (which again is different from the bad science described previously) on the overall scientific enterprise is miniscule: its incidence is very low, much of it is found in relatively obscure journals and uncited papers, and the self-correcting aspects of the enterprise are good at rooting it out. Much greater damage can be done to society at large, however; an example, the infamous Wakefield paper on the measles, mumps, and rubella vaccine, will be explored further in the next chapter. Furthermore, fraud damages the reputation of the scientific enterprise. It can be used as an excuse to dismiss any inconvenient scientific results and can lead to both a decrease in support for public research funding and a general rise in anti-science attitudes.

12
THE TRIUMPH OF MISINFORMATION; THE PERILS OF IGNORANCE

Human history becomes more and more a race between education and catastrophe.

—H. G. WELLS

MMR AND AUTISM

In 1998, the British physician Dr. Andrew Wakefield and twelve coauthors published a paper in the leading British medical journal *The Lancet* that reported on a study of twelve children (eleven boys and one girl) aged three to ten years old who all had developmental disorders (eight had autism) and had been referred to gastroenterologists for various abdominal symptoms. After a very extensive series of gastrointestinal and neurological tests, the authors discussed possible links between the children's developmental disorders and their gastrointestinal complaints. They also noted that the first symptoms of the developmental disorders were reported by parents between twenty-four hours and two weeks after the children received the measles, mumps, and rubella (MMR) vaccine. The one line in Wakefield and his coauthors' article that stood out particularly was their admission that they did not prove an association between the MMR vaccine and the syndrome described. Nonetheless, at a press conference Wakefield called to mark the article's release, he said, "It's a moral issue for me. I can't support the continued use of these three vaccines, given in combination, until this issue has been resolved."[1]

It turns out, however, that many other relevant points were not included in the paper. For example, Wakefield ordered highly invasive tests such as spinal taps and conducted the entire "study" without submitting it to an ethics review board as required; indeed, he didn't even have the medical authority to undertake such tests. However, by far the most disturbing fact not included was that, over the two years prior to the article's publication, Wakefield had been paid $720,000 by a team of lawyers that was attempting to generate a case against the MMR vaccine manufacturers—on behalf of the parents of some of the children in the study.

Indeed, Wakefield was not the only person making out handsomely from his totally fabricated claims. One of the referees of one of his articles was reportedly paid over $60,000. Five of his colleagues at North London's Royal Free Hospital where the study was done were also on the take to the tune of over $300,000. Furthermore, in June 1997, a year before the infamous article was published, Wakefield filed a patent for a "safer vaccine" meant to bring about "the elimination of MMR."[2]

Six years later, in February 2004, investigative journalist Brian Deer of the UK's Channel 4 and the *Times* broke the news of the Wakefield fraud. Ten of the coauthors of the original paper (not including Wakefield) issued a partial retraction of the first paper. In February 2010, *The Lancet* fully retracted the paper several days after Wakefield was stripped of his ability to practice medicine in the United Kingdom by the General Medical Counsel that, in a 143-page report,[3] called his behavior irresponsible and dishonest.

The impact of this fraud on public health was, and continues to be, enormous. Vaccination rates in the United Kingdom fell from 92 percent in 1996 to 78.9 percent in early 2003. As a consequence, the number of cases of mumps rose from 4,200 in 2003 to 56,400 by 2005. In April 2006, a thirteen-year-old boy who had not been vaccinated became the first person to die from measles in the United Kingdom in fourteen years. In 1994, the only cases of measles that occurred in the United Kingdom were imported by foreigners who had not been vaccinated. By 2008, the Stockholm-based European Center for Disease Prevention and Control reported that the disease was again endemic in the United Kingdom; halfway through the year, 530 cases had been reported. In 2012, there was a measles outbreak in two Welsh counties that led to 1,219 cases, eighty-eight hospitalizations, and one death.

Politicians, the media, and ignorant doctors all contributed to the hysteria. Prime Minister Tony Blair and his wife Cherie refused to tell the media whether they had had their son vaccinated; the fact that Cherie wore crystals to keep evil rays from harming her evidently did not undermine her credibility with the press and public. The Royal College of General Practitioners supplied ignorant anti-MMR doctors for press interviews. The media dredged up emotionally wrought parents who were told they had caused their children's autism by subjecting them to MMR inoculations.

U.S. Representative Dan Burton (a member of the Tea Party caucus whose early career was spent trying to impeach Bill Clinton, whom he called "a scumbag") turned his attention to the vaccine/autism controversy when his "only grandson became autistic right before [his] eyes—shortly after receiving his federally recommended and state-mandated vaccines." Burton had Dr. Wakefield testify before his Committee on Government Reform "investigating" the purported vaccine/autism link on June 19, 2002.

During his testimony, Wakefield claimed he had examined many more children since the 1998 article and that he had identified close to 150 more cases in which MMR vaccinations had led to autism.

According to the U.S. Centers for Disease Control and Prevention,[4] there are an estimated 164,000 measles deaths per year worldwide, mostly in countries without access to immunization programs. Prior to an effective measles vaccine, three to four million U.S. residents contracted the disease each year, leading to 48,000 hospitalizations, 1,000 chronic disabilities from encephalitis, and 500 deaths annually. Today, approximately 90 percent of children in the United States receive the MMR vaccine; in 2012, there were only fifty-five measles cases in the United States, a remarkable achievement brought about by scientific medicine and scientifically informed public health campaigns. In 2013, however, worrying signs appeared when cases jumped to nearly 200. As in previous years, all cases were originally imported, but instead of a visitor infecting no one or perhaps one other child, one visitor infected fifteen people at a Texas megachurch at which parents routinely refuse to vaccinate their children.[5]

Jenny McCarthy and her ilk would like to return us to this third-world situation. With a career launched on the basis of being *Playboy* magazine's Playmate of the Year (an obviously legitimate substitute for medical credentials), McCarthy has appeared in numerous movies and television shows and, in 2013, was appointed as co-host of the television talk show *The View* (she lasted less than a year before being fired). Her son was diagnosed with autism in 2005, and during an appearance on *Larry King Live* in 2008, she took up the Wakefield theory, which had been thoroughly debunked by then, and audaciously stated that vaccines can trigger autism. Since then she has used her celebrity to promote this highly irresponsible view in books (in which she claims "mommy instincts" are obviously superior to scientific medical evidence), on websites, and in a foreword to Wakefield's book *Callous Disregard: Autism and Vaccines—The Truth Behind a Tragedy*.

Many groups have been spawned to support Wakefield's fraud. As a consequence of such antivaccine (and antirationality) groups, hundreds of thousands of U.S. parents are now refusing to allow their children to be vaccinated. As the vaccination rate drops below the herd immunity threshold of roughly 90 percent, the number of cases of MMR will soar, just as they have in the United Kingdom.

If you conduct a Google search for Andrew Wakefield, the first page that comes up has three (out of twelve) references touting his thoroughly discredited vaccine/autism connection, including "New Published Study Verifies Andrew Wakefield's Research on Autism" (from the LibertyBeacon .com site, where one can also learn how the American government is waging a war on its own people by secretly engineering the California drought) and, from the *Huffington Post*, "*The Lancet* Retraction Changes Nothing." It doesn't matter that ten of his coauthors have disavowed the research, that the journal in which it was published has retracted the article, that several attempts to reproduce his results have failed completely to do so, that his license to practice medicine has been revoked because of his deliberate fraud, or that his blatant conflict of interest has been revealed and remains unchallenged. All of this is spun into a vast conspiracy among governments, vaccine producers, and the scientific establishment to punish this heroic figure. The Web gives free and equal access to all beliefs, however irrational. Seventeen years after his original publication, the story will not die. Only children will.

CANCER CLUSTERS

I am writing this chapter at my house on the end of the North Fork of Long Island, that hundred-mile-long dump of glacial deposits described in the first chapter. For decades a variety of groups have touted their belief that Long Islanders are in the midst of a breast cancer epidemic and that industrial and agricultural chemicals are responsible. The oldest group,

1 in 9: The Long Island Breast Cancer Action Coalition, even managed to get federal legislation passed in 1993 that mandated the National Cancer Institute conduct a large study to examine explicitly five possible environmental risk factors: contaminated drinking water, sources of indoor and ambient air pollution, electromagnetic fields, pesticides and other toxic chemicals, and hazardous and municipal waste.

Having a credible scientific organization study possible causes of cancer doesn't sound like a problem—unless one recognizes that it is based on a false premise, directs scarce resources in unfruitful directions, and generally replaces rigorous and rational scientific judgment with the "beliefs" of activists and politicians (in this case Senator Alfonse D'Amato) who, lacking the ability (and/or the will) to make informed judgments on their own, pander to them.

The putative Long Island epidemic is an example of the general case of "cancer clusters," the concept that in certain locations and at certain times, an unusually high number of cancers are detected. Real such clusters are worrisome, of course, and strongly suggest an environmental trigger. The problem, however, is that very few of the so-called clusters are real.

Typical of the "evidence" for cancer clusters is that cited by a Long Island woman in an interview with the Associated Press: the fact that three women on her block have died of cancer, she opined, "is not a coincidence. It can't be."[6] For readers who absorbed the lessons of chapter 6, I hope the reaction is, "Well, it can be, and it probably is."

Equipped with the Poisson distribution, we know how to calculate the probability of three deaths on her block from cancer if we know the average death rate from cancer. The American Cancer Society estimates (far too precisely given the accuracy of the input data) that we should have expected 585,720 cancer deaths in the United States in 2014. There are about 320 million people in the United States, so that gives the average death probability from cancer in a year of 1.83×10^{-3}. Now we don't know for sure how many people live on this woman's particular block, but if there are ten houses on each street, that's about forty houses with between two and three people per household, so let's make it 100 people (probably not off by a factor of two). That means the odds of one person dying of cancer on her block (depending on the age distribution and many other demographic factors, but we just estimating roughly here) is 100 × $1.83 \times 10^{-3} = 0.183$.

We also aren't told over what period of time these three women died, but let's make the extreme assumption it was all in one year. When the expected average is 0.183, and the outcome is 3, the Poisson distribution tells us the probability of this happening is, from chapter 7, $P(3) = (0.183)^3$ $e^{-0.183}/3! = 8.5 \times 10^{-4}$.

Now there are 2.85 million people living on Long Island—and that's from Nassau and Suffolk counties alone. (Because Queens and Brooklyn are a part of New York City, they are left out of the study despite the fact that, geographically speaking, they are unambiguously a part of Long Island.) That means there are 28,500 independent samples of 100 people. So we should expect this woman's experience to happen $8.5 \times 10^{-4} \times$ 28,500 = 24 times. Every year!

Suppose twice as many—six people—on her one block all died of cancer over a three-year period. That would be an obvious cancer cluster, right? No, we would expect it to happen once every few years on Long Island. The point is, as stated in chapter 6, rare things happen all the time. And they require no explanation.

Nonetheless, the large Long Island Breast Cancer Study Project, now mandated by law, went ahead with coordination and funding from the National Cancer Institute and with the collaboration of the National Institute of Environmental Health Sciences. Ten separate studies examined all of the potential environmental factors cited in the legislation, and extensive data for more than 1,500 women newly diagnosed with breast cancer and a roughly equal number of cancer-free controls were collected. The results of each study were published in the refereed literature. None of the environmental factors was found to have any association with breast cancer risk. On the other hand, the known risk factors such as family history of breast cancer, having a first child when older than twenty-eight, never having a child, age, and higher income were all confirmed among this sample. That is, absolutely nothing was added to our knowledge in exchange for the millions of dollars (and dozens of scientists' efforts) expended.

In July 2012, Goodman et al.[7] published a review of 428 cancer cluster investigations from throughout the United States. In only thirteen percent of the cases was an actual increased incidence of cancer confirmed, and in only one of the 428 putative cluster cases was the cause of cancer revealed. It is clear that public health resources are best spent in other ways. It is also clear that this won't happen as long as public policy is informed by irrational emotionalism rather than logical thought.

DISTURBING CONCLUSIONS

Space does not permit me to exhaust the list of irrational decisions and fantasy-based public policies that are to be found in our misinformation-saturated world. Jonathan Swift made a daunting observation for those of us who want to change the basis of public and private decision-making: "Reasoning will never make a man correct an ill opinion," he said, "which by reasoning he never acquired"[8] And it seems that the application of reason is increasingly succumbing to identity politics—if you identify with a political party, special interest group, or other social entity, you just adopt the "facts" that entity espouses.

A fascinating recent study of this phenomenon was recently reported by Dan Kahan of Yale Law School.[9] By testing general science knowledge as well as many specific facts about climate change, he found little difference between subjects that self-identified along the political spectrum from far left to far right; i.e., it is not a lack of scientific "knowledge" that climate change deniers exhibit. Even worse, he found that high-level reasoning abilities did not differ appreciably between the two groups. It is not, therefore, a problem of education. Rather, he posits,

"Every individual . . . employs her reasoning powers to apprehend what is known to science from two, parallel perspectives simultaneously: a *collective-knowledge-acquisition* one, and a *cultural-identity-protective* one."

He gives a dramatic example of this phenomenon from surveys conducted by the National Science Foundation. After measuring both "ordinary science intelligence" and "religiosity," a true/false question about evolution is posed in one of two forms. The first is as follows:

Human beings, as we know them today, developed from earlier species of animals. (True/False)

Those in the lowest 10 percent of science intelligence show no difference in their ability to answer correctly (about one-third say true), but for the rest of the population, the percentage of (scientifically) correct answers diverges dramatically. For those of above-average religiosity, a constant 30 percent say true irrespective of general science intelligence, whereas for the less religiously inclined even those with a mean science intelligence score answer correctly more than 80 percent of the time.

Check out what changes, however, when we alter the question to read as follows:

According to the theory of evolution, human beings, as we know them today, developed from earlier species of animals. (True/False)

A dramatically different result is found in this case. Those in the bottom 10 percent of science intelligence have a lower percentage of correct answers independent of religiosity (~20 percent), but for both groups, the fraction of correct answers rises rapidly such that those with the mean science intelligence score give the correct answers 80 and 90 percent of the time for above- and below-average religiosity, respectively. That is, most citizens, religious or not, know the scientific facts; one group simply rejects them out of hand to cleave to their social identity.

Kahan goes on to show with a nationally representative sample of 1,800 adults that climate change is an even more polarized issue than evolution. Again, knowledge of "what climate scientists believe" was quite similar among Democrats and Republicans, with 75 to 85 percent of both groups answering basic questions correctly (e.g., CO_2 is the worrisome greenhouse gas, human-caused global warming will flood coastlines, etc.) and more sophisticated questions incorrectly (e.g., the melting of the North Polar ice cap will result in sea level rise—a canard we debunked earlier). But when asked to rate the risks associated with global warming, those with the highest science comprehension scores diverged most in their assessments depending on their political affiliation, with Democrats ranking the risk 6.5/7 and Republicans ranking it 1.5/7.

Kahan concludes:

When "what do you believe?" about a societal risk validly measures "who are you?," or "whose side are you on?," identity-protective cognition is not a breakdown in individual reason but a form of it. Without question, this style of reasoning is collectively disastrous: the more proficiently it is exercised by the citizens of a culturally diverse democratic society, the less likely they are to converge on scientific evidence essential to protecting them from harm. But the predictable tragedy of this outcome does not counteract the incentive individuals face to use their reason for identity protection.

Collectively disastrous indeed.

13
THE UNFINISHED CATHEDRAL

WHAT WE KNOW

I had a student some years ago who was having difficulty in my class and came in during office hours one day for help.[1] The topic that particular week was the phases of the Moon. Now, I do not consider it essential for life as an intelligent citizen to understand why the Moon has phases. However, I include the topic in my course for two reasons. First, as noted earlier, most people cling to an incorrect model of why the Moon's phases occur, and the topic serves as a useful illustration of how it is sometimes necessary to unlearn incorrect knowledge before acquiring new knowledge. Second, it is a good example of a simple physical model against which one can compare reality—or so I thought.

After an hour with this student, having used my basketball Earth and tennis ball Moon to illustrate the three-dimensional essence of the problem, I finally asked her, "OK, so at what time of day would we expect to see the waning crescent moon overhead in the sky?"

"At 9 A.M.," she replied.
"Yes! Excellent, I think you've got it."
"Oh," she said, "but that must be wrong."

"No, you are correct!"

"No, no, that can't be right."

"Why not?" I asked.

"Because the Moon isn't up in the daytime."

"Yes, it is sometimes," I assured her.

"No it isn't."

"Well, as a matter of fact, if you come here tomorrow morning about 10
   A.M., I'll show it to you right outside my window."

Her parting response as she left my office: "I don't have to; I know the
Moon's not up in the daytime."

You may find this anecdote amusing. I find it frightening. The key word
here is *know*, meaning, of course, as the dictionary defines it, to be aware
through observation, inquiry, or information. In the Age of Misinformation,
I have a problem with the "or" in this definition. My student obviously had
observed the Moon in the sky—apparently always at night. She was comfortable enough with this anecdotal observation that she saw no need for
inquiry—or even for assimilating new information. Just personal, haphaz-
ard observation is viewed as sufficient for determining what she "knows." It
doesn't matter if she continues to insist that the Moon is only up at night and
that when it's full it makes people crazy; that the Earth is only 6,000 years old
and that this is the age she wants her children to be taught; that the measles,
mumps, and rubella vaccine causes autism; that global warming is a hoax. . . .

Or does it?

The postmodernist concept of equally valid, individually constructed
realities is not an arcane academic fad—it is a threat to our survival as an
advanced civilization.

HOW WE KNOW

The late cosmologist Ted Harrison divided the history of human views
of the universe into three eras: the anthropomorphic, anthropocentric,

and anthropometric.[2] The anthropomorphic universe is an Age of Magic. There is no boundary between the self and the external world; natural phenomena are regarded as manifestations of human emotions. Some of this persists in our language: we talk of angry storms and gentle breezes. We imagine this to be the cosmology of prehistory, although pockets of anthropomorphic thinking remain prevalent (in such places as Southern California and in certain university English departments). It is a primitive worldview fundamentally incompatible with life in a modern technological society.

The anthropocentric view finds expression in the Age of Myth, when a pantheon of powerful gods was created to explain natural phenomena, although it is noteworthy that these gods were driven by human emotions and obsessed with human concerns. It is, in a profound sense, an Earth-centered universe, and it remains the predominant view in most of the world today. U.S. Senator James Inhofe is exemplary of this kind of thinking: how arrogant of us to think we could actually change Earth's climate when God is in charge of that? The anthropocentric view renders impossible the management of a planet with seven billion intelligent creatures that each consume 100 times their metabolic energy requirements.

The anthropometric view represents the Age of Science. It is not, as Protagoras would have it, that man is the measure of all things, but it assumes that he can take measurements of an external reality. This age began when Copernicus removed the Earth from the center of the universe. It leads to a profoundly powerful model of Nature as well as an entirely new perspective on the cosmos and, more importantly, our place in it. And as I trust is clear by now, it is this last view to which I subscribe: (1) that there exists an objective, material universe of space, time, matter, and energy; (2) that we are fully, and I underscore *fully*, a part of this material universe; and (3) that our brains, having evolved an adaptive capacity of considerable power in the service of the survival and reproduction of our species, can, in their spare time, be put to work to understand this universe.

WHAT'S WORTH KNOWING

If while relaxing at the beach I asked you to show me a particle, you would most likely pick up a grain of sand—a small solid entity with a mass, a shape, and a color, sitting motionless on your fingertip. If I then asked you to show me a wave, you would point to the undulating motion of the ocean, the water moving rhythmically up and down until it breaks and the wave's energy is dissipated as the water is thrust up the beach. There is a clear and obvious distinction between the particle and the wave—a solid, stationary object versus a periodic motion carrying energy.

The world inside an atom, however, is very different from the one you routinely experience. Its constituent components do not recognize the clean distinction you make between particles and waves. In exploring the subatomic world, we can perform experiments on these components and measure how they behave. In some circumstances, their behavior is best described as that of a particle, whereas in different circumstances, they look for all the world like a wave. The same is true of light.

In describing this to students, I am often asked what the electron is really like, what its *true* characteristics are, whether it is a particle or a wave, what the *meaning* of this dual existence is, etc. My answer to these questions is rarely viewed as satisfactory: "I don't know, and I don't care."

I really don't. I cannot, ever, even in principle, experience life as an electron, so it is a waste of time worrying about what this would truly be like. I am comfortable with the notion that I will never be like an electron. But I am fascinated by how electrons behave in atoms—how their internal dance produces the light emerging from the fluorescent tube over my head, how their patterns determine the links they can form with neighbors, and how those links make a lemon sour and an orange sweet. So I use the tools of science to poke around inside atoms, gather data on their behavior, and build models of their reality. It's perfectly fine their reality is not like mine. But if my models work—and until I falsify them, they do—I can predict the sweetness of oranges and the sourness of lemons and control the color balance of my fluorescent light. I have gained a partial understanding of some natural phenomena and can use this understanding for my own pleasure and comfort. Who needs truth? Who needs meaning?

In chapter 1, I extolled the virtue of the child's curiosity and argued that it is one of the distinguishing features of humankind. Now we must ask: What form should the answer to a question such as why the sky is making loud noises take?

There's the anthropocentric answer: "Because God is angry with the sinners on Earth and is expressing his rage, warning them to repent." Such answers come from those seeking absolute truth.

There's the anthropomorphic answer: "Because the sky spirits are fighting, hurling rocks at each other, but they'll stop soon, and the air will be fresh with their sweet breath as they make up." Such answers come from those who seek meaning.

Finally, there is the anthropometric answer: "Because of the interaction among the air, water, and ice in a cloud, the bottom of the cloud tends to become negatively charged with respect to both the top of the cloud and the ground. When the difference in charge becomes great enough, the air can no longer hold the charges apart, and a current flows rapidly between them. This heats the air along the electrical discharge to a temperature of greater than 20,000 K. The superheated air expands very rapidly, creating a shock wave just like the sonic boom of a jet. This wave of sound travels outward in all directions, jiggling the air molecules back and forth until they reach your eardrums and make them vibrate in synchrony, allowing you to hear the sound of thunder." Such an answer comes from those who seek understanding.

The anthropometric answer is longer. It involves linking together many bits of hard-won knowledge. It presents a model consistent with that knowledge and subject to further testing. It connects the questioner's senses to an external stimulus and explains his or her experience. It sees the world as understandable.

Those who seek truth can abjure thinking altogether. They know everything is taken care of and beyond their control.

Those who seek meaning think of ways to connect their personal feelings and sensations to an anthropomorphic universe. They want to believe everything is taken care of and beyond their control.

Those who seek understanding recognize the remarkable analytical ability our brains provide and apply that power to seek explanations for the behavior of the material universe. With understanding comes a limited degree of control and a responsibility to utilize the knowledge gained in a manner consistent with physical constraints. With understanding comes a deep appreciation for the marvelous workings of the universe and the extraordinary kilogram of matter that allows us to comprehend it.

It is my view that we seek truth and meaning when we lack understanding. Understanding is much more satisfying, more useful, and more fun.

THE ESSENCE OF SCIENCE

Science is neither a collection of unquestioned facts nor a simple recipe for generating more facts. Rather, it is a process of inquiring about Nature, and given that Nature is not only much bigger than humans but has also been around a lot longer, it should come as no surprise that we haven't finished the job: there is no complete scientific description of the universe and all it contains. The gaps in our knowledge are what make science so engaging.

Science is an intensely creative activity, but it differs in several important ways from other creative human activities such as art, music, or writing. Science is teleological. It has a goal toward which it strives—an ever-more accurate and all-encompassing understanding of the universe. Artistic endeavors do not share such a goal. Sculptors since the time of Michelangelo have not been striving to improve upon the accuracy of his sculpture of David. Writers have not been searching for four hundred years to find arrangements of words more universal than "A rose by any other name would smell as sweet." Scientists have, however, been working steadily on Galileo's ideas of motion, expanding their scope and improving their accuracy to build a more comprehensive model of motion through space and time.

Science and art also differ in regard to the role of the individual. Artistic creativity largely expresses one person's vision. Contrary to some caricatures, science is a highly social activity. Scientists have lots of conferences, journals, websites, and coffee hours—they are constantly talking and writing, exchanging ideas, collaborating and competing to add another small tile to the mosaic of scientific understanding. It is from the conventions of this social web that the self-correcting nature of science emerges.

At any given moment, many "scientific" ideas are wrong. But as the last 400 years have shown, science as a whole has made tremendous progress. The reason for this progress is that wrong ideas in science never triumph in the end. Nature is always standing by as the arbiter and, while the aether may have survived for more than two millennia (chapter 9), as soon as the physical and mathematical tools were in place to measure its properties, its absence was readily discovered and accepted.

The enterprise of science has developed several habits and techniques for enhancing the pace of correcting false ideas. Perhaps foremost among these is skepticism. Although many people regard skepticism rather negatively, it is a scientist's best quality. Indeed, it is essential to be skeptical of one's data, always to look for ways in which a measurement might be biased or confounded by some external effect. It is even more essential to be skeptical of one's own models and to recognize them as temporary approximations that are likely to be superseded by better descriptions of Nature. When one's data agree with one's model, euphoria must be tempered by a thorough and skeptical critique of both.

When the results of one's experiment or observation are ready for public display, community skepticism springs into action. The description of one's work in a scientific publication must provide enough detail to allow other scientists to reproduce one's results—they're skeptical and want to make sure you did it right. The published article itself is the result of a formal skeptical review by one or more referees chosen by journal editors for their expertise in a certain field and their objective, critical, skeptical approach. These days, instant publication of results via the Internet often precedes formal publication in a journal, exposing the author to dozens or hundreds of unsolicited, skeptical reviews from those who scan new postings each morning.

All this skepticism screens out a lot of nonsense right at the start. Furthermore, it optimizes the ability of both the original author and other scientists to root out errors and advance understanding. The constant communication through both formal and informal means rapidly disseminates new ideas so they can be woven quickly into the fabric of our current models, offering further opportunities to find inconsistencies and to eliminate them. This highly social enterprise with this highly skeptical ethos is central to the rapid growth of scientific understanding in the modern era.

My celebration of skepticism, emphasis on falsifiability, and insistence on the temporary nature of models should not be misconstrued as supporting the popular notion that science consists of an endless series of eureka moments. The news media are committed devotees of this false view. Each week, the science section of the *New York Times* needs half a dozen "news" stories, and if they can use words like "stunning," "revolutionary," and "theory overturned" in the headlines, so much the better. Scientists are complicit in this misrepresentation, all too easily using phrases such as "breakthrough," "astonishing," and the like, not only when talking to reporters but even when writing grant proposals and journal articles. Some philosophers of science share the blame by concentrating their studies on "paradigm shifts" and "scientific revolutions." Science isn't really like that.

Science is much more like building a cathedral than blowing one up. Thousands of hands place the stones, weave the tapestries, tile the frescos, and assemble the stained-glass windows. Occasionally, a new idea might require the disassembly of some work already completed—invention of the flying buttress allowed the walls to go higher, so a new roof was needed. Very infrequently, on timescales typically measured in centuries, a genuinely new conception of the cathedral's architecture emerges. While a major supporting wall or facade may need to be removed, we use many of the stones again, rehang some of the old tapestries, and always enclose most of the old building within the new. Our cathedral gets larger and ever more ecumenical, drawing a greater swath of the universe within its doors as the weaving, the tiling, and the stonemasonry goes on. It is extraordinarily gratifying and important work.

APPENDIX

Practicing Scientific Habits of Mind

CHAPTER 3 PRACTICE:
SCALES, MEASUREMENTS, AND BIG NUMBERS

I expect that many readers have learned about the metric system, basic geometry, scientific notation, and unit conversions. I also expect that for many, this was a long time ago. Furthermore, if you learned these things in the way they are often taught, it is unlikely you have internalized them to the point where you can use them effortlessly in solving problems of the type you will encounter in the Misinformation Age. I provide the following set of problems so that you may determine whether your skillset is rusty and needs some polishing before delving into the more advanced topics I will cover in later chapters. The answers can be found below.

1. The Central Park Reservoir is surrounded by a well-worn jogging path approximately 1 km in length and the reservoir is 20 m deep. How many cubic meters of water does it contain?

2. The metric system is designed so that water links its units of length, mass, and volume: $1 \text{ cm}^3 = 1$ milliliter $= 1$ g. What is the total mass of the water in the reservoir (in kilograms)?

3. If the hydrogen in the H2O has a mass of one unit, the oxygen a mass of sixteen units, and one (atomic mass) unit $= 1.67 \times 10^{-27}$ kg, how many hydrogen atoms are there in the reservoir?

4. If we could recreate the conditions at the center of the Sun on Earth, we could use these hydrogen atoms to make energy by sticking four of them together to make helium (with the added benefit of cheaper helium balloons). For each helium atom produced, we would get 2.5×10^{-12} joules of energy (it takes 100 joules each second to power a 100-watt light bulb). How much energy would be produced if I converted all the hydrogen in the reservoir to helium?

5. Total energy use in the United States is now roughly 8.9×10^{19} joule/year. How long would the reservoir meet our energy supply? (You'll understand after answering this one why engineers are so anxious to make controlled fusion reactors work.)

ANSWERS

1. $1.6 \times 10^6 \text{ m}^3$
2. 1.6×10^9 kg
3. 1.1×10^{35} atoms
4. 6.9×10^{22} joules
5. 770 years

CHAPTER 4 PRACTICE: USING ENVELOPES FOR FUN AND PROFIT

As chapter 4 notes, solving back-of-the-envelope problems requires a logical approach coupled with a dose of self-confidence. Such confidence comes with practice.

The news media and Hollywood never seem to tire of stories about killer asteroids and comets. Collisions of solar system detritus with Earth are, in fact, not infrequent, and even civilization-ending ones are not unknown, but that last big one was 65 million years ago, so it's not something I'd lose a lot of sleep over. But suppose you were asked by one of your less scientifically literate relatives the following question: "What are the odds of my house being struck by a meteor during my lifetime?" If I gave you the one critical piece of information you may very well not know—that there are about 100,000 objects a year softball-sized or larger that strike the Earth—could you quickly calculate a rough estimate?

If the answer is yes, you have probably mastered this habit of mind and you need read no further. If you haven't a clue as to how to proceed, or won't have any confidence in the answer you might obtain, read on.

A logical approach to these (or any other) problems that I have found effective is given here.

I. UNDERSTANDING THE PROBLEM

(a) Write down what you are looking for. It is absolutely essential to know where you are going if you hope to get there—include units; units are critical because they clarify exactly what you want to know and will provide an extremely useful check that you have done the problem correctly.

(b) Write down in the simplest form possible all the information you are given. Again, include units.

(c) Draw a picture if appropriate—this often helps you to visualize what you have, and what you need to get an answer.

II. IDENTIFYING ADDITIONAL INFORMATION

(a) Write down things you may need to know and the units in which you wish to know them.

(b) Write down, in mathematical symbols if appropriate, the definitions of key words and phrases.

(c) Don't get sidetracked by peripheral issues or minor possible corrections; remember, the point here is to get a *rough* estimate of some quantity.

(d) Be explicit about the assumptions you are making.

III. MAKING GOOD ESTIMATES

(a) Estimate the things you don't know. This is obviously a key point and may not be easy at first. One trick is to start with either absurdly large or absurdly small values that you know must be wrong and try to converge from these outrageous values to something more reasonable.

(b) Think about whether your assumptions are justifiable.

IV. GETTING THE ANSWER

(a) Calculate, keeping careful track of units and factors of ten; convert units when necessary (see chapter 3 exercises).

(b) Do a sanity check on your answer. This is not always straightforward—after all, if you knew roughly what the answer should be you wouldn't have had to do the calculation. But you can at least check that the units of the answer are right (e.g., if you were trying to estimate the number of people doing something and got an answer in gallons, there's probably a problem). You might also have at least some clue about the size of the answer—e.g., if you were trying to estimate the number of high school teachers in New York City and got a number larger than seven billion (the world's population) you can tell there's a problem. Below I provide some worked examples, followed by a half dozen for you to try. The answers can be found at the end.

* * *

Sample problem 1: On September 23, 2014, the New York City subway system set a single-day record for the number of riders with 6.1 million passengers; the average that month was just under five million per day. Roughly how many gallons of gas were saved by the subway system that month (the power for the subway is largely hydroelectric power from Quebec)?

I. UNDERSTANDING THE PROBLEM

(a)   What we want to know: gallons of gas saved in a month = gallons per month.

(b)   What we are given: passengers per day = 5×10^6.

(c)   Picture probably not required.

II. IDENTIFYING ADDITIONAL INFORMATION

(a)   What we need to know: average trip length per rider; number of passengers per car if passengers were driving instead; miles per gallon for the average car in city traffic.

(b)   Definitions: all pretty straightforward.

(c)   There are probably as many peripheral issues that could present stumbling blocks as there are people who attempt this problem. For example, what about all those tour groups of thirty people (or more) who would otherwise all be on one bus? What fraction of subway riders do you think they make up? I'd guess less than 1 percent—and we certainly aren't worried about corrections at that level.

III. MAKING GOOD ESTIMATES

(a)   Average trip length: There are twenty New York City blocks per mile, so a ride from Columbia to Greenwich Village is a little over five miles. A commute from the outer boroughs can be more than ten; a short trip to the Metropolitan Opera is only 2.5. My estimate is five miles per trip. A factor of ten smaller would be under a mile—most New Yorkers would just walk. A factor of ten longer would be fifty miles, which is more than the distance from the northern Bronx to southern Brooklyn (a trip I suspect few people make regularly). Note how testing much bigger and much smaller numbers gives us added confidence in our estimate.

(b)   Number of passengers per car: I drive as little as possible (which is one of the main reasons I like living in New York City), but when I do, it looks to me as though most people are alone in their cars. Perhaps the subways have a few couples or family groups traveling together, but I suspect most subway commutes are likely to be solitary. I'd estimate the average is somewhere between one and two—let's take 1.5.

(c)   Miles per gallon for the average car: those of you who do drive probably know this better than I do, but I think the average fuel efficiency of U.S. cars is around twenty-five miles per gallon. Substituting for the subway, of course, would mean heavy city driving (indeed, it would be heavy city traffic jams if three to four million more cars came to New York each day!), so that's probably an overestimate. I'll adopt twenty miles per gallon.

IV. GETTING THE ANSWER

(a)   All right—we're ready to calculate: 5×10^6 riders/day × 1 car/1.5 riders × 30 days/month × 5 miles/car × 1 gallon/20 miles = 2.5×10^7 gallons or 25 million gallons in one month! Note how all the other units canceled out and how we were left with what we wanted: gallons per month. If we had mistakenly put in miles per gallon instead of gallons per mile in that term, the answer would have had the wrong units, cluing us in to the likelihood we made a mistake. Note also that we quote our answer to two significant figures since some of the input numbers are only accurate to 20 percent (or even less). See chapter 7 for more than you may want to know about significant figures.

(b)   Is this number plausible? Sure—there are almost five million people involved, and the idea of a person using five gallons of gas in a month isn't crazy (most use more, but not 100 times more or 100 times less).

Sample problem 2: How about those meteors? Could your house really be hit by one?

I. UNDERSTANDING THE PROBLEM

(a)   What we want to know: the odds of one's house being hit in a lifetime.

(b)   What we are given: 10^5 meteors big enough to reach the Earth's surface and penetrate a house's roof arrive each year.

(c)   Picture: probably not useful here. Although you might want to think about the relative size of a softball and a house roof. The latter is obviously much bigger, so the softball will either hit the roof or it won't, and we only need to worry about the size of all targets (roofs). If we were dealing with 10-km-wide asteroids (such as the one that wiped out the dinosaurs), it's much bigger than a roof, and we'd have to worry about its size as well. Fortunately, we only have to worry about ones such as that every 100 million years, so we can probably ignore it in this instance.

II. IDENTIFYING ADDITIONAL INFORMATION

(a) What we need to know: we need to calculate the fraction of the Earth's surface covered by roofs to get the fraction of the time a roof is hit. This will require knowing the Earth's population and estimating the average roof area per person, as well as the size of the Earth. The Earth's population is a number you should remember: $\sim 7 \times 10^9$. The radius of the Earth is easy to look up, but it's also something you should be able to roughly estimate (can you?): it's $\sim 6,400$ km.

(b) Definitions: a basic one is what I mean by "probability." It is simply the number of outcomes of interest (in this case, a roof getting hit) divided by the total possible outcomes (anywhere on Earth getting hit)—see chapter 6 for an extended discussion of probability. We'll also need to know how to calculate the area of a sphere ($4\pi R^2$) and the area of a roof (length × width).

(c) One peripheral issue might be all the people living in apartment buildings in cities. They have much less roof per person. This may be true, but (1) slightly over half the world's population lives in rural areas; (2) many people living in cities live in normal houses in the first world or in shanty towns in the third world; and (3) the estimate still doesn't change much (e.g., there are forty-three apartments—roughly 100 people—living in my Columbia apartment building, and it's roof area is about 900 m^2, so that would be 9 m^2, which, you will see, is within a factor of two or so of my rough estimate).

(d) Assumptions: I am assuming meteors land randomly all over the Earth; since I can't think of any reason they would have preferred landing places, this seems plausible.

III. MAKING GOOD ESTIMATES

(a)   Roof area per person: here it is important to recognize that most of the world's population does not live the American suburban lifestyle, so don't say 100 m^2 (which would be the answer for a four-person family in a 4,000-sq. ft. house). My estimate would be 4 m^2—and even that may be generous. My reasoning is that most of Earth's population does sleep indoors. The area needed for an adult to sleep is a minimum of 1 m^2, so it has to be bigger than this. On the other hand, it is almost certainly not ten times this area because that would mean each person having a 10 × 10-ft. living space, and that is certainly an overestimate given that roughly 40 percent of the world's population has an income of less than $2 per day.

IV. GETTING THE ANSWER

(a)   Ready to calculate: 7×10^9 people $\times$ 4 m^2/person$/4\pi(6{,}400$ km $\times 10^3$ m$/1$ km$)^2 =$ 5×10^{-5} is the fraction of the Earth's surface covered by roofs. There are 100,000 chances per year, suggesting this happens several times a year somewhere in the world ($5 \times 10^{-5} \times 10^5$/yr $= 5$/yr). In fact, it happened in 2003 just outside Chicago (see www.meteorobs.org/maillist/msg27439.html).

But what about *your* house? Just because someone wins the lottery every day doesn't mean you will. U.S. average household size is 2.6 people, and life expectancy is approaching eighty years. In addition, our average roof area per person is, as noted previously, probably almost a factor of ten above the world average. So if the world average is five times per year there would be 400 hits in your lifetime, and a U.S. roof makes up 40 m^2 × 2.6 people/roof$/4\pi(6{,}400$ km $\times$ 10^3m$/1$ km$)^2 = 2 \times 10^{-13}$ of the Earth's surface $\times 10^5$ per year $\times$ 80 years $= 1.6 \times 10^{-6}$, or a little over one a million—better than the odds in most state lotteries but not something you should lose sleep over.

(b)   Is this plausible? Hard to say offhand, but a modest amount of Internet research shows that there are several reported incidents of meteors that have crashed through roofs in the United States in the past few decades, and since we make up only 4.5 percent of the world's population, that's about what one would expect if it happens several times a year somewhere in the world.

Here are some practice problems for you to try:

1. If you lived in a town with a population of 25,000 people, roughly how many people would have the same birthday as you (feel free to substitute the actual population of your hometown)? Approximately how many have the same day *and* date (i.e., were born on exactly the same day and year)?

2. McDonalds now claims they have sold more than 300 billion burgers. If you were to spread these all out in layers over an area the size of Columbia's main campus, would you be up to your knees in burgers? The main campus is six blocks long and three blocks wide (see section 2c for a translation of blocks to distance). Suppose you made a single stack—how high would it be (ignoring the fact that the weight of the stack would squeeze all the grease out)?

3. Estimate the number of dump trucks filled with the toenail clippings are generated each week on Earth.

4. The diameter of an atom is about 10^{-10} m (they're all about the same size). Cells have an enormous range of sizes—from tiny red blood cells to some neurons in your spinal cord that have extensions from a single cell more than two feet long. But let's take a more typical cell—it is about 0.5 micron (1 μm = 10^{-6} m) in diameter and is roughly (good enough for this calculation) spherical. How many atoms are there in such a cell?

5. Calculate the approximate number of atoms in the period at the end of this sentence.

6. The Greenland ice sheet extends roughly 1,700 km from north to south and has an average width of 1,000 km. Its mean thickness is about 1,500 m, although in some places it is more than 3,000 m thick. If it were all to melt and run off into the world's oceans that cover about 70 percent of the planet today, how much would sea level rise?

ANSWERS

Here are the answers for the six practice problems. If you are within a factor of two or three on numbers 1, 4, 5, and 6, and within a factor of ten or twenty on the others, you're on your way to mastering this scientific habit of mind.

1. Roughly seventy with the same birthday and maybe one with an identical birthdate.

2. The layer would be more than 1500 meters high—*way* over your knees. A single stack would reach roughly thirteen million miles high or over fifty times the distance to the Moon.

3. A few, I'd say.

4. About 100 billion atoms in a cell—same as the number of stars in the Milky Way, the number of galaxies in the visible Universe, and the number of neurons in your brain (and, yes, that is all a coincidence).

5. About 10^{18} atoms, or a billion billion—periods are a lot bigger than most cells.

6. Since sea level rise from melting ice caps is an interesting number for you to know, I solve this one in detail here:

What we want to know: the amount of sea level rise in meters.

What we are given: the ice is 1,700 km × 1,000 km × 1,500 m, the area of the oceans is 70 percent of the total area of Earth; the Earth's radius is 6,400 km (from the previous meteor example). Note we are also given the glacier's greatest thickness, but this will not be relevant. Be sure to ignore irrelevant information.

Picture: if it is unclear how the volume of ice is going to spread out and produce a change in the average depth of the ocean change, a picture might help—with this giant ice cube melting into a thin surface layer of water, it should be clear that the volume of the ice is to be compared to the volume of the new ocean = area of ocean × added depth.

What we need to know: volume of ice, area of Earth's surface; volume = length × width × height; surface area of a sphere = $4\pi R^2$.

Estimates: we don't need any here—we have everything, unless you want to worry about the fact that ice is somewhat less dense than water and therefore takes up more space (it floats). But first, most of this ice is compressed under many tons of ice above it and so isn't less dense; furthermore, this is less than a 10 percent effect, and most of our numbers aren't that accurate, so we can ignore this.

Ready to calculate: volume of ice = 1,700 km × 10³ m/km × 1,000 km × 10³ m/km × 1,500 m = 2.55 × 10¹⁵ m³. Note how we converted all the units to meters (we could have converted them all to kilometers, but since our answer is likely to be in meters we chose that). Now since we realized from the picture that the volume of water is just the area of the ocean's surface times the depth, all we need to do is to divide this volume by the area. The ocean's surface = 4π(6,400 km × 10³ m/km)² × 0.7 = 3.6 × 10¹⁴ m². Dividing this into the ice volume, that answer is about 7 m or about 23 ft. (which is not good news for most of the major cities of the world).

CHAPTER 5 PRACTICE: READING AND APPRECIATING GRAPHS

The following questions all refer to graphs in chapter 5 and are referenced by their figure numbers.

1. In figure 5.2, what are the two frequencies at which the intensity is 200 MJy/sr (megajanskys per steradian, if you're interested)?

2. In figure 5.13, what is the most common range in which to find the daily high temperature? How many days a year does this occur? Be careful!

3. In the bottom plot of figure 5.16, which point is the biggest outlier from the trend?

4. In figure 5.21, which is the steeper approach to Pike's Peak (the closed contour roughly halfway between points A and B)—from A or from B?

5. In figure 5.23, are there any objects outside the dashed box whose limits are consistent with them being inside the box?

6. In figure 5.25, what is the approximate uncertainty in the position of the giant black hole at the center of our galaxy?

7. In figure 5.30, which represents a quasar putting out more energy, one at a de-reddened absolute K magnitude of –30 or –32? What is the maximum redshift at which we could detect a K = –30 quasar with a reddening of $E(B - V) = 0.75$?

ANSWERS

1. Roughly 2 and 10 (1/cm)
2. 70 to 80°F; it occurs about seventy-eight days per year (since this is the data from two years, you must divide the *y* value by two— remember, read the caption!)
3. BCH
4. A
5. No
6. About ±0.01 in each coordinate
7. −32; at about redshift 1.2

CHAPTERS 6, 7, AND 8 PRACTICE: PROBABILITY AND STATISTICS

1. You have two dice. You roll one and get a six.
   (a) What are the odds of getting a six when you roll the other one?
   (b) Suppose you roll them simultaneously; what are the odds of rolling a twelve?
   (c) What are the odds of rolling an eleven?

2. If you calculate a correlation coefficient $r = 0.5$ with a sample size of twenty-two, what is the probability, quoted to two significant figures, that the correlation happens by chance?

3. With a full deck of fifty-two cards, what are the odds of holding a jack, a queen, and a king after three consecutive draws?

4. What are the odds of rolling a pair of sixes if you roll two dice ten times? Try using the binomial distribution. What about getting two pairs of sixes in ten tries?

5. You are conducting an experiment to control the corn borer (a worm that infests corn crops and can cause billions of dollars of damage to the U.S. corn crop) and have discovered a natural compound that seems to kill them in the lab.

To see if this new treatment works in practice, however, you need to spray it on a field of infested corn. You expect to find an average of twenty borers per ear of corn without any control. You march into the field a couple of days after the first treatment, and in the first ear of corn you pull you find only eight worms. Should you be excited (i.e., what is the probability of getting eight when you expect twenty—try the Poisson distribution)?

6. If you set a radioactive nucleus with a half-life of one minute on your desk, what is the probability that it will decay between now and one minute from now? If it is still there five minutes from now, what is the probability that it will decay between the fifth and sixth minute from now?

ANSWERS

(1a) 1/6; (1b) 1/36; (1c) 2/36 or 1/18

(2) 1.9 percent (or 0.019)

(3) $12/52 \times 8/51 \times 4/50 = 2.9 \times 10^{-3}$, or a little over 1/4 of 1 percent

(4) 0.27 of one success and 0.028 for two

(5) Yes, since the odds of getting eight when you expect twenty is 0.0013, or about one-tenth of 1 percent. Now, if there are hundreds of ears of corn and you just happened to get the one with only eight borers, it still could mean your new treatment is useless, but if your next ear only has eight as well, your hopes would be justified in soaring because, as you no doubt instantly recognized, the probability of getting eight and eight when you expected twenty and twenty is 0.0013 × 0.0013, or only two in a million.

(6) It is, and in any given minute will always be, 50 percent. Do not buy into the gambler's fallacy.

NOTES

INTRODUCTION: INFORMATION, MISINFORMATION, AND OUR PLANET'S FUTURE

1. IBM. 2015. "What Is Big Data?" Available at: www-01.ibm.com/software/data /bigdata/what-is-big-data.html. Accessed January 17, 2015.
2. E. O. Wilson. 2012. *The Social Conquest of Earth*. New York: Liveright.

1. A WALK IN THE PARK

1. Although Manhattan streets are labeled "East" and "West," they're actually rotated such that the avenues run 29 degrees east of north; this means the cross streets actually run southeast and northwest. For an entertaining disquisition on this subject by Dr. Neil deGrasse Tyson, see www.amnh.org/our-research /hayden-planetarium/resources/manhattanhenge.
2. Physicists are not, in general, tristadecaphobes (people fearful of the number thirteen); unlike most hotels in New York, Pupin Physics Laboratory at Columbia University proudly has an elevator button for the thirteenth floor.
3. Richard Louv coined the term *nature-deficit disorder* in his book *Last Child in the Woods: Saving Our Children from Nature-Deficit Disorder* (Chapel Hill, NC: Algonquin Books, 2005).
4. Because the Earth travels around the Sun in an elliptical path, the distance sunlight has to travel to reach the Earth is constantly changing. The average

separation is 149,600,000 kilometers (or, to be exact—and we'll get to what that means in chapter 8—149,597,870.7 km). Because light travels 300,000 kilometers per second (or, to be exact, 299,792.458 km/sec), it takes 499.00 seconds (or eight minutes and nineteen seconds) for the light to reach Earth's surface on average. Over the course of a year, this time varies from eight minutes and ten seconds to eight minutes and twenty-seven seconds.

5. Keats, "Lamia," part 2, lines 229–239. See www.bartleby.com/126/37.html.

6. Margaret Robertson. 1909. "Life of Keats," p. 202. In *Poems Published in 1820*. Gloucestershire, UK: Clarendon Press. See https://ebooks.adelaide.edu.au/k/keats/john/life-of-keats.

7. Edgar Allen Poe, "Sonnet—To Science." See http://www.poetryfoundation.org/poem/178351.

8. "U.S. and World Population Clock," *United States Census Bureau*, see www.census.gov/popclock/.

9. R. B. Alley, 2013, "Watchable Wildlife and Demand-Driven General Education." *Journal of General Education* 62(1): 39.

10. For a bee's-eye view of a crocus flower, see Klaus Schmitt, "Spring Crocus—Cerco 94mm Quartz Fluorite lens for reflected ultraviolet photography," *Photography of the Invisible World* (blog), March 6, 2013; see http://photographyoftheinvisible-world.blogspot.com/2013/03/spring-crocus-cerco-94mm-quartz.html.

2. WHAT IS SCIENCE?

1. John Hawks. 2011. "Selection for Smaller Brains in Holocene Human Evolution," *John Hawks* (blog); see http://johnhawks.net/research/hawks-2011-brain-size-selection-holocene and references therein.

2. It is important to note in this regard, however, that even axiomatic mathematical systems cannot find everything that's true. Kurt Gödel's first incompleteness theorem demonstrates that in any consistent formal system there are statements expressible in the language of that system that cannot be proven either true or false.

3. Karl Popper. 1978. "Science, Pseudoscience, and Falsifiability." In *Conjectures and Refutations* (pp. 33–39). London: Routledge.

4. S. Haack. 2013. "Six Signs of Scientism." *Skeptical Inquirer* 37(6): 40.

5. Popper. 1992. *Unended Quest*, sec. 33. New York: Routledge.

6. Jonathan Weiner. 1994. *The Beak of the Finch*. New York: Knopf.

7. In Greek mythology, Aeolus was the ruler of the winds who kept the various wind spirits locked away until he was commanded by the gods to release them, unleashing storms and snatching people and things from Earth for Zeus.

8. Note, however, that we still do give human names to hurricanes and cyclones, although most of us do not impute their destructive power to angry gods. The National Oceanographic and Atmospheric Administration has an entire website devoted to the rationale behind naming tropical storms, although it is silent as to why strong storms at higher latitudes remain nameless.

9. Andrew Read. 2013. "Science in General Education." *Journal of General Education* 62(1): 28–36.

10. The first of the planets unknown to the ancients and invisible to the naked eye is Uranus, and its discovery in 1781 by William Hershel created a sensation. By the 1840s, Uranus had nearly completed one orbit of the Sun, and it was recognized that its path varied slightly from that predicted by Kepler's laws of planetary motion, a variation that could be explained by the existence of yet a more distant, massive planet. In 1845, Joseph Le Verrier of the Paris Observatory and John Couch Adams of the University of Cambridge each set out to calculate the expected location of this new planet using Newton's law of gravitation. On September 23, 1846, Johann Gottfried Galle of the Berlin Observatory pointed his telescope to the location Le Verrier had predicted and discovered Neptune. This was regarded as both a spectacular confirmation of Newton's universal law of gravitation and a triumph for the scientific observation of Nature.

11. Jacob Bronowski. 1956. *Science and Human Values.* New York: Messner.

12. Keith Stanovich. 2004. *The Robot's Rebellion.* Chicago: University of Chicago Press.

13. Carl Sagan. 1997. *The Demon-Haunted World: Science as a Candle in the Dark.* New York: Ballantine.

3. A SENSE OF SCALE

1. Christiaan Huygens. 1690. *New Conjectures Concerning the Planetary Worlds, Their Inhabitants and Productions.* London: Printed for Timothy Childe in 1698.

2. The antipodes of any point on the Earth is the point opposite it connected by a straight line passing through the planet's center. In other words, it is the farthest distance you can reach when traveling over the Earth's surface.

3. The mean distance of the Moon from the Earth is 384,391 km, and the distance to Proxima Centauri, the nearest star to the Sun on this scale, would be 762,000 km.

4. There are three different temperature scales in general use, but only one of them is based on what we now understand to be the physical meaning of temperature: a measure of the mean energy of motion of the atoms and molecules that

make up a substance. A sensible scale would then have a value of zero when all motion stopped. That is the definition of the Kelvin scale used here. The intervals (degrees) on the Kelvin scale are the same as those on the centigrade (or Celsius) scale: one one-hundredth of the distance between the freezing and boiling points of water at the surface of the Earth. This makes the zero point $-273.16°C$. Thus, $5,780$ K $= 5,780 - 273 = 5507°C = 9945°F$. By convention, one uses "Kelvins" rather than "degrees Kelvin."

5. I have taken the fourth movement of Beethoven's Ninth Symphony ("Ode to Joy") and imposed a series of one-octave-wide filters up and down the scale to "blind" your ear to all but that single octave, just as your eye selects only a single octave of the radiation the Universe sends us and blinds us to the rest. The demonstration can be downloaded at http://ccnmtl.columbia.edu/projects/helfand.

6. A radian is just another unit for measuring angles. Just as a circle can be divided into 360 degrees, it an also be divided into 2π radians. Thus, 1 radian $= 360/2\pi$ $= 57.3$ degrees.

7. Note that what we have done here is convert the units so they cancel out. That is, we first multiply the width of the hair in millimeters by the number of meters in a millimeter and then divide by the distance in meters. This gives us an answer with no units, since it is an angle (measured, as noted, in radians).

8. Scientific notation is used in many areas of science because the scales of interest are often very large or very small compared with the human-sized units we use in everyday life. The general form of a number in scientific notation is $N \times 10^x$, which, in other words, reads as follows: N times 10 raised to the power of x, where x is called the *exponent* or *power of 10*. To convert an ordinary number to its scientific notation counterpart, there's a two-step process.

First, locate the decimal point of your number and move it either to the right or to the left so that there is only one nonzero digit to the left of it. This new number is your N.

Second, count the number of spaces you just had to move your decimal point. If you moved to the left, then each space adds 1 to x (where x begins at zero). If you moved to the right, then each space adds -1 to x.

EXAMPLE 1: 5,230,400. For step 1, you need to move the decimal point six places to the left. The number now reads 5.230400, or 5.2304 (since any trailing zeroes may be dropped). So 5.2304 is your value of N. For step 2, recall that you have moved six places to the left so that your value is $x = 6$. So your number in scientific notation is 5.2304×10^6, or 5.2304 times 10 to the sixth power.

EXAMPLE 2: 0.00038. This time, you move the decimal point to the right four spaces so that $N = 3.8$ and $x = -4$. So your number in scientific notation becomes 3.8×10^{-4}, or 3.8 times 10 to the negative fourth power.

9. The Sun-like star Alpha Centauri is actually the brightest star in a triple star system and, at the moment, the closest star is the junior member of this triplet, Proxima Centauri, whose orbit has carried it just under 3 percent closer to Earth than Alpha.

10. Ade, P.A.R. et al. 2014. "Planck 2013 Results, XVI. Cosmological Parameters" *Astronomy & Astrophysics*, 571: 16P.

INTERLUDE 1: NUMBERS

1. This was true as of June 7, 2015. The results will, of course, change with time as the dynamic software of Google constantly revises which of these websites is hottest at the moment—who knows, Angels may be number 1 by now.

2. Galileo Galilei. 1623. *The Assayer*. (*Il Saggiatore*)

3. "Inflated Applicants: Attribution Errors in Performance Evaluation by Professionals," S.A. Swift, D.A. Moore, Z.S. Sharek, & F. Gino PLOS ONE, Published: July 24, 2013. DOI: 10.1371/journal.pone.0069258. http://journals.plos .org/plosone/article?id=10.1371/journal.pone.0069258

   Wall Street Journal article 31 July 2013 by M. Korn, http://www.wsj.com /articles/SB10001424127887323997004578640241102477584

4. Coral Davenport. 2014. "White House Announces Climate Change Initiatives." *New York Times*, July 16, 2014. http://www.nytimes.com/2014/07/17/us/politics /white-house-unveils-climate-change-initiatives.html?_r=0

5. Sigma Xi, "Your World of Science News." https://www.smartbrief.com/servlet /ArchiveServlet?issueid=85B862DC-C67A-4C85-B163-25E77 DEB8382&lmid=archives.

6. Mark Brown. 2013. "Harry Winston's Turning Point: Can the miner get Ekati?" *Canadian Business*, January 21, 2013. www.canadianbusiness.com/companies -and-industries/harry-winston-turning-point.

4. DISCOVERIES ON THE BACK OF AN ENVELOPE

1. Mercury, like the other planets, traces an elliptical orbit as it travels around the sun—at least approximately. The ellipse is not fixed. It gradually rotates so that Mercury's point of closest approach to the Sun—called the perihelion of the orbit—drifts, or precesses, about 0.0015 degrees per year. Why the precession? According to Newton's theory of gravity, any two massive bodies attract each other. If we want to apply this "classical" view of gravity to Mercury's motion, we

must take into account the effect not only of the Sun, which primarily determines Mercury's orbit, but also of other bodies relatively close to Mercury; it turns out that Venus, Earth, Jupiter, and the other planets all contribute non-negligible perturbations. Calculating the combined effect, we find that our theoretical calculation for Mercury's rate of precession is 7 percent off from the measured value, a discrepancy in the observed position of Mercury in the sky that accumulates over the course of a century to an error equivalent to the width of a human hair observed one meter away—a truly tiny effect, but one that the French astronomer Urbain Le Verrier deduced in 1859 by carefully studying records of Mercury's motion over the preceding 150 years. In his general theory of relativity, Einstein presented an entirely new view of gravity as the manifestation of the interplay between massive bodies and space itself. He suggested that a mass doesn't merely float in an unperturbed vacuum—it actually *shapes* the space around it, warps it, curves it. In this picture, a mass experiencing the pull of gravity is simply following its natural path through a curved space. A star warps space the same way a heavy ball will warp a rubber sheet—it creates a depression. Nearby objects (e.g., planets) follow trajectories in this dented space (i.e., their "orbits") because their motions are determined by the shape of the space through which they are moving. By applying this interpretation to the problem of Mercury's orbital precession, the 7 percent discrepancy vanishes entirely. Einstein first showed this was approximately true in a paper published in late 1915, during which time he first presented his general theory. In a letter to his friend Paul Ehrenfest, he wrote, "Imagine my delight that . . . the equations yield Mercury's perihelion motion correctly. I was beside myself with joy and excitement for days." An exact derivation was completed and reported to Einstein by Karl Schwarzschild in a letter dated December 22, 1915. Schwarzschild concludes: "It is an entirely wonderful thing, that from one so abstract an idea comes out such a conclusive clarification of the Mercury anomaly." And indeed it is. General relativity was not created to solve the Mercury anomaly problem. It arose from Einstein's deep thoughts about the very nature of space and time, matter and gravity. The tiny discrepancy in Mercury's orbit had been known for more than half a century. In this case, the high degree of accuracy with which Mercury's motion had been measured was important enough to overthrow Newton's theory of gravity.

2. US Census Bureau. www.census.gov/popclock.

3. CDC. 2014. "Tobacco-related mortality." www.cdc.gov/tobacco/data_statistics /fact_sheets/health_effects/tobacco_related_mortality.

4. Data from the National Highway and Transportation Safety Administration, which reported 32,719 deaths in 2013, the last year from which full statistics are available.

5. *New York Times.* 2002. "Thousands Forced to Evacuate as Fires Rage Across Colorado." www.nytimes.com/2002/06/10/us/thousands-forced-to-evacuate-as-fires-rage-across-colorado.html

6. Opinion. 1997. "Safety standards for imported food." *The New York Times*, October 3.

7. www.npr.org/2012/12/24/167977420/why-we-behave-so-oddly-in-elevators

8. Matt Richtel and Brian Chen. 2014. "Tim Cook, making Apple his own." *The New York Times*, June 15.

9. US Census Bureau. www.census.gov/popclock.

10. The Common Era, or c.e., refers to the most frequently used calendar system: the Gregorian Calendar. It has the same definition and value as Anno Domini (Year of our Lord in Latin), or a.d., which corresponds to the approximate birth year of Jesus Christ. To steer clearly of religious connotations and the likely inaccuracy of the historical record, scientists now replace a.d. with c.e., adopting as the arbitrary starting point for this enumeration of years the date of January 1, 01 c.e. decreed by the Gregorian Calendrical reform of 1582. For earlier times, we use b.c.e.—Before the Common Era.

11. The Hubble Ultra-Deep Field, the deepest picture of the Universe every taken, is described in detail at www.stsci.edu/hst/udf.

12. My way of visualizing a trillion is to imagine a warehouse covering a full city block and twenty-five stories tall. Now fill it with marbles—it takes about a trillion marbles.

13. On August 6, 1945, the United States dropped an atomic bomb over the city of Hiroshima, Japan. Japan refused to surrender, and three days later (August 9), a second bomb was dropped over Nagasaki. By August 14, Japan had surrendered unconditionally, and World War II had ended.

14. Child Trends Databank. 2014. "Motor vehicle deaths." www.childtrends.org/?indicators=motor-vehicle-deaths.

5. INSIGHTS IN LINES AND DOTS

1. Rick White, private communication.

2. When light enters a medium such as air, it can scatter off the particles that make up the medium. The degree of scattering depends of the wavelength, or color, of the light. In our atmosphere, the shorter wavelengths (blue and violet) scatter more effectively than the longer wavelengths (orange and red). (Scattering off particles much smaller than the wavelength of the light such as air molecules is called Rayleigh scattering and depends on the inverse fourth power

of the wavelength, meaning that violet light scatters ten times more efficiently than the red light.) This is why a clear sky is blue—the red, orange, yellow, and green wavelengths make it straight through (and combine to show us a whitish-yellow sun), whereas the blue wavelengths scatter all over the place such that they appear to come from all directions. Near sunset, however, all the light must pass through much more of the Earth's atmosphere, and even the green and yellow light gets scattered around, producing the orange-red sun we see just before it dips below the horizon. The addition of small particles to the atmosphere (tiny water droplets, volcanic dust, soot, etc.) enhances the effect.

3. When authors complete a manuscript describing their work and wish to disseminate it, they often submit it to a journal for publication. The journal editor sends the paper to one or more "referees" who are experts in the field to review, asking them to comment (often anonymously, sometimes not) on the originality, presentation, and scientific content of the paper and recommend revisions or rejection. This process is part of the healthy skepticism that is a hallmark of science—see chapter 13.

4. The phrase "as a function of" is one that scientists and mathematicians use routinely, but it can use some explanation for the uninitiated. A function defines one variable in terms of another. The statement "y is a function of x" (denoted $y = y(x)$) means that y varies according to whatever value x takes on. A causal relationship is sometimes implied (i.e., x causes y), but does *not* necessarily exist. For a university student, the number of classes the student is taking may be a function of his or her major. It is also likely to be a function of how many extracurricular activities the student participates in, how much sleep the student needs, etc. Knowing that one variable somehow changes with some specific attribute of a system is a starting point. Uncovering the specific nature of the relationship between that variable of interest and the system's attribute is another. We have developed an efficient way to represent such problems using the concept of a function. If $y(x) = 2x$, that means that for every incremental increase in x, y increases by two increments. For every value x takes on, y is twice its value. If $x = 5$, then $y = 2 \times 5 = 10$. If we say we are plotting one variable as a function of another, we are doing so to look for any relationship there might be between them. For a two-dimensional plot, each point much have at least two numbers associated with it, and where it is located horizontally on the plot depends on (is a function of) its x value.

5. I had 731 data points—one for each day for two years: 365 + 366 = 731 (1964 had 366 days because it was a leap year).

6. If the possible values ranged from 17.5°F to 99°F, in increments of 0.5, you'll end up with $(99 - 17.5) \times 2 + 1 = 164$ discrete points. The +1 is because we have a bin for

17.5 and another for 99, whereas the difference × 2 does not count both ends. So the number of whole numbers from 5 through 10 could be expressed as follows: 6 = (10 − 5) + 1.

7. S.-J. Leslie, A. Cimipian, M. Meyer, and E. Freeland. 2015. "Expectations of Brilliance Underlie Gender Distributions Across Academic Disciplines." *Science* 347: 262–265.

8. Figure 4 in E. Glikman, M. D. Gregg, M. Lacy, D.J. Helfand, R.H. Becker, & R.L. White 2004. "FIRST-2Mass Sources Below the APM Detection Threshold: A Population of Highly Reddened Quasars." *Astrophysical Journal* 607: 60–75.

9. Edward Tufte, a professor emeritus of political science and statistics at Yale University, has written seven books and delivered countless lectures on the subject of the graphical representation of data. His book *The Visual Display of Quantitative Information* (Cheshire, CT: Graphics Press, 1983) devotes an entire chapter to "graphical integrity" (i.e., avoiding misrepresentations). For more on Tufte, see www.edwardtufte.com/tufte.

10. The Chandra Deep Field-North survey. www2.astro.psu.edu/users/niel/hdf /hdf-chandra.html.

11. Figure 12 in "The FIRST-2MASS Red Quasar Survey," Glikman, E., Helfand, D.J., White, R.L., Becker, R.H., Gregg, M.D., & Lacy, M2007. *Astrophysical Journal*, 667: 673.

INTERLUDE 2: LOGIC AND LANGUAGE

1. Coral Davenport. 2014. "Rising Seas," *New York Times*, March 28, 2014. www .nytimes.com/interactive/2014/03/27/world/climate-rising-seas.html.

6. EXPECTING THE IMPROBABLE

1. The New York Metropolitan Transportation Authority, which runs the subway system, estimates that 35 percent of these deaths were suicides. http://www.nydailynews .com/new-york/nyc-subway-train-deaths-decrease-2013-mta-article-1.1562928

2. This may sound like a lot of people, but there were 1.71 billion riders in 2013, meaning there are $53/1.71 \times 10^9 = 3.1 \times 10^{-8}$ deaths per trip or odds of 1 in 32 million. In contrast, in 2012, there were 33,561 highway fatalities. Since there are 253 million registered vehicles of all types on the nation's highways, and we can estimate that each, on average, makes two trips per day, the corresponding number

for the odds of dying in any one trip in a year are 33,561 deaths per year/2 trips per day per vehicle × 253 × 10^6 vehicles × 365 days/year or 1.8 × 10^{-7} deaths per trip or odds of 1 in 5.5 million, nearly six times higher than riding the subway.

3. Centers for Disease Control and Prevention. http://www.cdc.gov/nchs/data /dvs/provisional_tables/Provisional_Table01_2013Dec.pdf.

4. Jeanne Meister, "Job Hopping Is the 'New Normal' for Millennials: Three Ways to Prevent a Human Resource Nightmare," *Forbes*, August 14, 2012. http://www .forbes.com/sites/jeannemeister/2012/08/14/job-hopping-is-the-new-normal -for-millennials-three-ways-to-prevent-a-human-resource-nightmare.

5. If I flip a coin a hundred times, I have ninety-one chances of getting ten heads in a row: 1, 2, 3, 4, 5, 6, 7, 8, 9, 10 and 2, 3, 4, 5, 6, 7, 8, 9, 10, 11 and . . . 91, 92, 93, 94, 95, 96, 97, 98, 99, 100.

6. J. A. Paulos. 1988. *Innumeracy: Mathematical Illiteracy and Its Consequences*. New York: Hill and Wang.

7. "United States Commercial Casino Gaming: Monthly Revenues," July 2015. University of Nevada Las Vegas Center for Gaming Research http://gaming .unlv.edu/reports/national_monthly.pdf.

8. Robert K. Merton. "The Self-Fulfilling Prophecy." *The Antioch Review* 8(2): 193–210.

9. T. Gilovich, A. Tversky, and R. Vallone. 1985. "The Hot Hand in Basketball: On the Misperception of Random Sequences." *Cognitive Psychology* 3(17): 295–314.

7. LIES, DAMNED LIES, AND STATISTICS

1. BBC World Service, "Mad on the Moon," *BBC News*, January 9, 2001. http://news .bbc.co.uk/2/hi/uk_news/1106830.stm.

2. The lunar phase calculator can be found at www.fourmilab.ch/cgi-bin/uncgi /Earth/action?opt=-m&img=Moon.evif.

3. The absolute value of a number is simply the value of the number with no plus or minus sign; it is indicated by enclosing a number or expression in vertical brackets: $|-7| = 7$. Thus, the absolute values of 7 and −7 are both 7. In this case, the distance in time between the full Moon and the shootings is all we care about. It doesn't matter whether the event occurred before or after the full Moon; we are only concerned with how far from the full Moon they happened.

4. In addition to the average (or mean) of a set of numbers, two other statistical terms are useful in describing the distribution: median and mode. The median is the number for which half of the other numbers are below this value and half are above; the mode is the number that occurs most frequently in the distribution.

Thus, in a sequence of thirty integers (3, 8, 7, 4, 5, 6, 3, 1, 6, 6, 7, 5, 4, 3, 6, 6, 5, 3, 4, 6, 7, 5, 3, 2, 5, 9, 5, 4, 6, 3), the mean is the sum (147) divided by 30 = 4.9; the median is 5 (there are 12 digits above 5 and 12 below it); and the mode is 6 (since there are seven 6's, more than any other digit). It is not always the case that the mean and median are similar. For example, if you listed the values of eleven cars in a parking lot that contained a bunch of subcompacts and a 2015 Rolls Royce Ghost (list price $321,900), it might look like this (in thousands of dollars): 15, 5, 11, 22, 12, 18, 7, 10, 14, 19, 322. In that case, the mean is $41,400, but the median of $14,000 is a much better description of the typical car in the lot.

5. This experiment is often attributed to Galileo, who is said to have dropped weights off the Leaning Tower of Pisa. There is, however, no documentary evidence that Galileo ever performed such an experiment. He did investigate the problem of falling bodies with different masses by rolling balls down inclined planes and independently concluded that the mass of an object does not affect the rate it falls under the influence of gravity. (See E. R. Harrison. 1981. *Cosmology: The Science of the Universe.* Cambridge, UK: Cambridge University Press, p. 125.)

6. These rules are adapted from https://quizlet.com/54025870/chemistry-flash -cards/.

7. Here are three examples of rounding numbers correctly:

(a) Round 4.3127 to four significant figures. Since 4.3127 has five significant figures, we need to drop the final 7, which leaves us with 4.312. But following rule 1 above, we note that since 7 is greater than 5, we need to add 1 to the last number we're keeping (the 2). So our rounded number becomes 4.313. You can appreciate that 4.313 is a better approximation of 4.3127 than 4.312 would be. Since the 7 we dropped indicates that the true value of the number is "closer" to 4.313 than it is to 4.312, we've created a better approximation by making that change.

(b) Round 10.417 to three significant figures. Since 10.417 has five significant figures we'll get rid of the last two. That leaves 10.4. The digit to the right of the 4 is 1—and 1 is less than 5. So we leave the 4 alone. Our final estimate is 10.4.

(c) Round 14.65 to three significant figures. Here the digit we want to get rid of is a 5. To decide what to do with the preceding digit 6, recall rule #3. If the number before the 5 is even, we leave it alone, and since 6 is even, our rounded value is 14.6.

8. The standard deviation of a distribution is determined by finding the average of the distribution, calculating the square of the difference between each number and this average, summing these squares, dividing by the number of points in the distribution (i.e., taking the average of the squared differences from the average),

and then taking the square root. As an example, the scores of a dozen students on a test might be as follows:

88, 81, 69, 94, 98, 59, 84, 80, 73, 90, 92, 82. The average of these twelve numbers is 82.5; the sum of the squares of the differences from this average is $[(82.5 - 88)^2 + (82.5 - 81)^2 + \ldots]/12 = 126$; and the square root of 126 = 11.2, and that's the standard deviation of this distribution (see chapter 7 on using standard deviations).

8. CORRELATION, CAUSATION . . . CONFUSION AND CLARITY

1. Suppose our sample had n = 22 and r = 0.467, neither of which exactly matches values shown in table 8.1. In this case, we use such a table by "interpolating" between the listed values. There are elaborate schemes for interpolation, but for most purposes, a simple linear interpolation is sufficient. Thus, for n = 22 and r = 0.467, we look at rows 20 and 25 and columns 0.4 and 0.5. This provides a little square (or matrix) of numbers:

$$r_0 = 0.4 \quad 0.5$$

n		
20:	0.081	0.025
25:	0.048	0.011

To interpolate the rows linearly, we take the difference between each pair of numbers, multiply by 2/5 (the fractional distance our number 22 is between rows 20 and 25), and then subtract the result from the row 20 value, yielding

$$r_0 = 0.4 \quad 0.5$$

n		
22:	0.0678	0.0194

We can then interpolate in the other direction and find that the probability of obtaining a linear correlation coefficient as large as 0.467 in a sample of twenty-two data points is 0.0354 or about 3.54 percent. In other words, once in every twenty-eight times (1/0.0354) we looked at a dataset of this size, we would find an apparent correlation this good when no relationship whatsoever existed between the quantities involved. It should be emphasized that this interpolated value is not strictly correct because the function that produces the table is not a linear one; i.e., the difference between 0.3 and 0.4 is not the same as the difference between 0.4 and 0.5, as you can readily verify for yourself by looking at the table. Over small intervals, however, a linear interpolation is usually sufficiently accurate.

9. DEFINITIONAL FEATURES OF SCIENCE

1. The *Oxford English Dictionary* defines "interregnum" as follows:

   (1) Temporary authority or rule exercised during a vacancy of the throne or a suspension of the usual government. (2) The interval between the close of a king's reign and the accession of his successor; any period during which a state is left without a ruler or with a merely provisional government. Bill Clinton was governor of Arkansas from 1979 to 1981 and then, after losing one election, from 1983 to 1992 when he became president.

2. Founded in 1963, the Creation Research Society proposed "to reach all people with the vital message of the scientific and historical truth about creation." In 1970, a splinter group formed the Creation Science Research Center in San Diego with the aim of reaching "the 63 million children of the United States with the scientific teaching of Biblical creationism." C.C. Young & M.A. Largent. 2007. *Evolution and Creationism: A Documentary and Reference Guide*, Westport, CT: Greenwood Press, 228.

3. H.M. Morris and M.E. Clark. 1976. *The Bible Has the Answer*. Green Forest, AR: Master Books, 31.

4. Most cases in federal courts bear little resemblance to television courtroom dramas. The vast majority of the work developing testimony and evidence is done offsite. In particular, most witnesses are deposed (questioned) at length, and the written transcript of the deposition is then submitted as evidence. It was this process in which I was involved.

5. From the decision by U.S. District Court Judge William Overton in the case of *McLean v. Arkansas Board of Education*, issued January 5, 1982. The full decision is available at www.talkorigins.org/faqs/mclean-v-arkansas.html.

6. Edwin Hubble used a class of pulsating stars called Cepheids to establish the extragalactic distance scale in the 1920s. It turned out, however, that the Cepheid stars in our galaxy that Hubble used to calibrate the distance scale were different from those in the distant galaxies he measured. This led him to adopt a distance scale that was too small by a factor of two and to the absurd notion that the Universe was younger than the Earth. The solution to this problem was found by Walter Baade in the early 1950s. A succinct history can be found at the American Institute of Physics' history of cosmology website www.aip.org/history/cosmology/ideas/hubble-distance-double.htm.

10. APPLYING SCIENTIFIC HABITS OF MIND TO EARTH'S FUTURE

1. Emmarie Huetteman, "Rubio on a Presidential Bid, and Climate Change," *New York Times*, May 11, 2014, www.nytimes.com/2014/05/12/us/politics/rubio-says -he-is-ready-to-be-president.html.

2. Here's the youtube clip: www.youtube.com/watch?v=EKd6UJPghUs. Here's one of dozens of citations: www.rightwingwatch.org/content/james-inhofe -says-bible-refutes-climate-change.

3. Coral Davenport, "Climate Change Deemed to Be Growing Threat by Military Researchers," *New York Times*, May 13, 2014, www.nytimes.com/2014/05/14 /us/politics/climate-change-deemed-growing-security-threat-by-military -researchers.html.

4. Alon Harish, "New Law in North Carolina Bans Latest Scientific Predictions of Sea-Level Rise," *ABC News,* August 2, 2012, http://abcnews.go.com/US /north-carolina-bans-latest-science-rising-sea-level/story?id=16913782.

5. In this plot, the uncertainties in the individual measurements are smaller than the size of the points, so no error bars are required. The line simply connects the points and does not represent a model of any kind.

6. You might realize that the opposite cycle must persist in the Southern Hemisphere because the seasons are reversed there (and that's correct), but there is so much more land area in the Northern Hemisphere (just look at a globe) that the growing season in this hemisphere dominates the cycle.

7. Data from WeatherSpark at https://weatherspark.com/averages/28802/12 /Amsterdam-Noord-Holland-The-Netherlands. Monthly values were obtained by replacing the 12 with each month's number (January = 1, February = 2, etc.).

8. "Working Group I Contribution to the IPCC Fifth Assessment Report (Ar5), Climate Change 2013: The Physical Science Basis," p TS-27. There is a slight awkwardness here, in that this report, while posted on the IPCC website, has the caveat that the report "should not be cited, quoted, or distributed", and the Working Group 1 (Physical Science Basis) final report does not include this result. Now, indeed, there are several legitimate possibilities as to why the models failed to predict the slowdown in temperature rise over the period 1998–2012 and this document both lists them—(a) internal climate variability, (b) missing or incorrect radiative forcing, and (c) model response error—and goes on to assess them. The problem arises when climate change deniers seize on this discrepancy as a reason to ignore all climate change science as they have—see "Climate Models' Tendency to Simulate Too Much Warming and the IPCC's Attempt to Cover That Up", by P. C. Knappenberger and P. J. Michaels (October 10, 2013). www.cato.org/blog /climate-models-tendency-simulate-too-much-warming-ipccs-attempt-cover.

11. WHAT ISN'T SCIENCE

1. Lynda Payne, "Health in England (16th–18th c.)," in Children and Youth in History, Item #166, http://chnm.gmu.edu/cyh/primary-sources/166 (accessed August 4, 2015).

2. "Medicine and Health." www.stratfordhall.org/educational-resources/teacher -resources/medicine-health/ (accessed August 4, 2015).

3. www.beliefnet.com/columnists/ohmystars/2014/08/the-astrology-of-robin -williams-behind-the-mask-of-comedy.html (accessed October 2, 2015).

4. Laura Donnelly, "Tory MP Says Astrology Is Good for the Health," *Telegraph*, July 25, 2014, www.telegraph.co.uk/news/health/news/10991455/Tory-MP-says -astrology-is-good-for-the-health.html.

5. S. Carlson. 1985. "A Double-Blind Test of Astrology." *Nature* 318: 419–425.

6. National Institutes of Health. 2013. "Homeopathy: An Introduction." https:// nccih.nih.gov/sites/nccam.nih.gov/files/Backgrounder_Homeopathy _05-23-2013.pdf (updated May 2013; accessed August 4, 2015).

7. NIH. 2014. "Homeopathy: An Introduction." https://nccih.nih.gov/health /homeopathy.

8. Mayo clinic. "Tests and Procedures: Acupuncture. Results." www.mayoclinic.org /tests-procedures/acupuncture/basics/results/prc-20020778 (accessed August 4, 2015)

9. C. M. Witt et al. 2009. "Safety of Acupuncture: Results of a Prospective Observational Study with 229,230 Patients and Introduction of a Medical Information and Consent Form." *Forsch Komplementmed* 16: 91–97. doi: 10.1159/000209315. Retrievable at www.ncbi.nlm.nih.gov/pubmed/19420954.

10. E. Ernst, S. L. Myeong, and C. Tae-Young. 2011. "Acupuncture: Does It Alleviate Pain and Are There Serious Risks? A Review of Reviews." *Pain* 152: 755–764. doi:10.1016/j.pain.2010.11.004.

11. Bruce Kennedy, "As Medical Costs Rise, More Americans Turn to Acupuncture," *Daily Finance*, April 14, 2011, www.dailyfinance.com/2011/04/02 /as-medical-costs-rise-more-americans-turn-to-acupuncture/.

12. Daryl Bem. March 2011. "Feeling the future: experimental evidence for anomalous retroactive influences on cognition and affect." *Journal of personality and social psychology* 100 (3): 407–425. doi:10.1037/a0021524. PMID 21280961.

13. David Helfand, "ESP, and the Assault on Rationality," *New York Times*, January 6, 2011, www.nytimes.com/roomfordebate/2011/01/06/the-esp-study-when-science -goes-psychic/esp-and-the-assault-on-rationality?module=Search&mabReward d=relbias%3Ar%2C%7B%221%22%3A%22RI%3A6%22%7D.

14. G. C. Begley and L. M. Ellis. 2012. "Drug Development: Raise Standards for Preclinical Cancer Research." *Nature* 483: 531–533. doi:10.1038/483531a.

15. Florian Prinz, Thomas Schlange, & Khusru Asadullah. 2011. "Believe It Or Not: How Much Can We Rely on Published Data on Potential Drug Targets?" *Nature Reviews Drug Discovery* 10(712).

16. Begley and Ellis, "Drug Development," 533.

17. C. F. Fang, R. G. Steen, and A. Casadevall. 2012. "Misconduct Accounts for the Majority of Retracted Scientific Publications." *Proceedings of the National Academy of Sciences USA* 109: 17028–17033. doi: 10.1073/pnas.1212247109.

18. M. Grieneisen and M. Zhang. 2012. "A Comprehensive Survey of Retracted Articles from the Scholarly Literature." *PLoS ONE* 7: e44118. doi: 10.1371/journal.pone.0044118.

19. What the abstract does not include is the fact that hundreds of these are retracted for technical reasons such as a journal publishing a wrong or duplicate issue and other such grounds that have nothing to do with fraud.

20. Ed Silverman, "Should Research Fraud Be a Crime? A Reader Poll," *Wall Street Journal*, July 16, 2014, http://blogs.wsj.com/pharmalot/2014/07/16/should-research-fraud-be-a-crime-a-reader-poll/.

12. THE TRIUMPH OF MISINFORMATION; THE PERILS OF IGNORANCE

1. Brian Deer. 2004. "MMR: The Truth Behind the Crisis." www.thesundaytimes.co.uk/sto/news/uk_news/Health/mmr/article31221.ece.

2. BrianDeer.com. http://briandeer.com/wakefield/vaccine-patent.htm.

3. BrianDeer.com. http://briandeer.com/solved/gmc-charge-sheet.pdf.

4. CDC. 2014. "Measles Vaccination." www.cdc.gov/measles/vaccination.html.

5. Liz Szabo, "Measles Cases in U.S. Rise; Most Unvaccinated, CDC Says," *USA Today*, December 5, 2013, www.usatoday.com/story/news/nation/2013/12/05/measles-cdc-vaccine-vaccinations-disease/3878375.

6. Medscape.com. www.medscape.com/viewarticle/442554.

7. M. Goodman, J. S. Naiman, D. Goodman, and J. S. LaKind. 2012. "Cancer Clusters in the USA: What Do the Last Twenty Years of State and Federal Investigations Tell Us?" *Critical Reviews in Toxicology* 42: 474–490.

8. J. Swift. 1720. "Letter to a young clergyman," available at *http://www.online-literature.com/swift/religion-church-vol-one/7/*

9. D. Kahan. 2015. "Climate Science Communication and the Measurement Problem." *Advances in Political Psychology Issue Supplement S1* 36: 1–43.

13. THE UNFINISHED CATHEDRAL

1. The first section of this chapter is adapted from my chapter "I'm Not a Heretic, I'm a Pagan," in *Neurosciences and Free Will*, edited by Robert Pollack (New York: Columbia University Press), p. 22.

2. E. R. Harrison, 1981. *Cosmology: The Science of the Universe*. Cambridge: Cambridge University Press, p. 13.

INDEX